SpringerBriefs in Criminology

SpringerBriefs in Policing

SpringerBriefs in Criminology present concise summaries of cutting edge research across the fields of Criminology and Criminal Justice. It publishes small but impactful volumes of between 50-125 pages, with a clearly defined focus. The series covers a broad range of Criminology research from experimental design and methods, to brief reports and regional studies, to policy-related applications.

The scope of the series spans the whole field of Criminology and Criminal Justice, with an aim to be on the leading edge and continue to advance research. The series will be international and cross-disciplinary, including a broad array of topics, including juvenile delinquency, policing, crime prevention, terrorism research, crime and place, quantitative methods, experimental research in criminology, research design and analysis, forensic science, crime prevention, victimology, criminal justice systems, psychology of law, and explanations for criminal behavior.

SpringerBriefs in Criminology will be of interest to a broad range of researchers and practitioners working in Criminology and Criminal Justice Research and in related academic fields such as Sociology, Psychology, Public Health, Economics and Political Science.

Series Editor
M. R. Haberfeld
John Jay College of Criminal Justice
City University of New York
New York, USA

SpringerBriefs in Policing presents concise summaries of cutting edge research in Police Science, across the fields of Criminology, Criminal Justice, Psychology, Forensic Science, and Corrections with implications for the study of police and police work. It publishes small but impactful volumes of between 50-125 pages, with a clearly defined focus. The series covers a broad range of Policing research: from experimental design and methods, to brief reports and regional case studies, to policy-related applications. The scope of the series spans the subfield of Policing, with an aim to be on the leading edge and continue to advance research. The series is international and cross-disciplinary, including a broad array of topics. The main goal of the series is to present innovations in Policing, in order to further the field as a research and evidence-based profession rather than a vocational occupation. It will showcase how Policing confronts problems and challenges that transcend cultures and borders and can be addressed from a global rather than local perspective. SpringerBriefs in Policing is aimed at a broad range of researchers and practitioners working in Criminology and Criminal Justice Research and in related academic fields such as Public Policy, Sociology, Psychology, Public Health, Economics, Policy Analysis, Terrorism and Political Science.

Anna Matczak

Adapting to Climate Change in Modern Policing

The Rise of the "Eco-Cop"

 Springer

Anna Matczak
The Hague University of Applied Sciences
The Hague, The Netherlands

This work was supported by Hague University of Applied Sciences (2800028).

ISSN 2192-8533   ISSN 2192-8541   (electronic)
SpringerBriefs in Criminology
ISBN 978-3-031-97509-7   ISBN 978-3-031-97510-3   (eBook)
https://doi.org/10.1007/978-3-031-97510-3

This Springer imprint is published by the registered company Springer Nature Switzerland AG
The registered company address is: Gewerbestrasse 11, 6330 Cham, Switzerland

If disposing of this product, please recycle the paper.

Preface

The twenty-first century has confronted police and policing with a series of profound challenges, but none as existential and far reaching as climate change. This book emerges from the accelerating climate crisis and from a recognition that climate change is not a distant concern for law enforcement; it is an immediate institutional, operational, and ethical reality.

As rising temperatures and sea levels, extreme weather events, and resource scarcity reshape the ways we live, work, and relate to one another, law enforcement agencies are being drawn into new and uncharted territory. For decades, climate change has been framed primarily as an environmental and scientific issue. Yet its cascading consequences now demand urgent attention from sectors traditionally seen as external to climate and environmental discourse. Among them, police and policing occupy a particularly complex position. On the one hand, law enforcement is tasked with maintaining order and continuity amid the disruptions caused by climate-induced displacement, disasters, resource conflict, and civil unrest. On the other hand, it must grapple with its own environmental and carbon footprint, operational adaptability, and legitimacy in an era of growing ecological awareness.

This book draws on the concept of climatisation—a framework for understanding how police services, often without explicit acknowledgement, are evolving and adapting in response to climate shocks, stressors, and ecological instability. It invites readers to think critically about the transformation of policing roles, the trade-offs inherent in such transformations, and the gradual emergence of the "eco-cop": a reimagined policing role and figure operating at the intersection of climate, environmental responsibility, and public safety.

The book is written for readers who seek not only to understand the topic, but also to foster the creation of supportive environments for the emergence and effectiveness of eco-cops. In addition to engaging with interdisciplinary literature and current theoretical debates, I spoke with 23 experts of diverse backgrounds and international profiles, situated in different parts of the world—individuals unlikely to meet at the same academic event, yet whose expertise offers distinct and complementary perspectives on this complex and multifaceted topic. Each conversation helped me to carefully examine the blurred boundaries between climate change and police work. These insights were then shaped into eight chapters—eight stories—through which I analyse the shifting terrain of modern policing as it is currently understood and practised.

In writing this work, I hope to contribute to a timely and necessary dialogue: one that does not treat climate change as an adjunct to criminology, policing studies, and police work but as a central, generative force in reshaping how we understand all three.

Competing Interests   The author has no competing interests to declare that are relevant to the content of this manuscript.

The Hague, The Netherlands   Anna Matczak

Acknowledgements

This book could not have come to fruition without the generous support, intellectual companionship, and encouragement of many individuals, to whom I extend my deepest gratitude.

First and foremost, I am profoundly indebted to Professor Maria Haberfeld, whose encouragement following my presentation at the 2023 European Society of Criminology conference in Florence was instrumental in shaping the direction of this project. I am equally grateful to Anna Goodlet and the editorial team at Springer for their professionalism, steady guidance, and unwavering support throughout the development of this manuscript.

My sincere thanks go to my colleagues and students at the Safety & Security Management Programme at The Hague University of Applied Sciences, as well as to the members of the Multilevel Regulation Research Group. I am especially thankful to Professor Barbara Warwas for creating a supportive and collegial environment, which supported the development of this book.

I wish to express my appreciation to the many experts who generously shared their time and insights during our conversations. These exchanges were not only intellectually rewarding but also among the most enriching experiences of this journey.

To my mum whom I owe heartfelt thanks for her constant encouragement and faith. Her enduring love and community-minded worldview continue to guide and sustain me in all that I do.

Finally, I dedicate this book, with deep affection and admiration, to my husband, Dr. Klaas Voss. His wisdom, unwavering faith in me, and our countless dinner-table reflections were the quiet force behind this endeavour, carrying it forward through many weekends, evenings, and holidays.

Contents

Chapter 1
Introduction

What Is This Book About?

The connection between climate change, police, and policing is counterintuitive to many, as the link between the three is not immediately apparent. Throughout the research on this topic, a range of reactions were encountered, including scepticism and curiosity. Some have asked, 'Does this mean that police officers will distribute water bottles on the street?' Others have inquired, 'Is this primarily about policing climate protests?' or even, 'Does it suggest that police officers need to adopt a vegan lifestyle?' These diverse responses only reinforced my conviction that there is a need for a comprehensive examination of this subject. This book seeks to provide a nuanced understanding of how climate change is reshaping police work, both by altering the nature of crimes that law enforcement must respond to and by driving structural and operational transformations within police organisations.

This book aims to explore the various processes, practices, and events through which the accelerating effects of climate change are transforming, and will continue to transform, police institutions and policing practices worldwide. Over the course of 3 years researching this topic, I frequently encountered the assertion that police services are largely unaware of this interrelationship and the broader debates surrounding it. At the same time, I observed that policing is rarely included in discussions about climate change, even in highly relevant and connected debates on climate and security.

In criminology, research on this topic has predominantly focused on the criminalisation of environmental and climate activism, often neglecting other critical dimensions of the relationship between police, policing, and climate change. However, as a frontline public safety service that operates in close and near-constant proximity to tensions, conflicts, harms, and crimes, the police should be regarded as a key actor in these discussions. Over the past 3 years, it has become increasingly clear to me that we are witnessing the formative years of 'green, eco-conscious, and climate-aware' policing. While a few pioneering examples exist, efforts remain fragmented and lack

A. Matczak, *Adapting to Climate Change in Modern Policing*, SpringerBriefs in Criminology, https://doi.org/10.1007/978-3-031-97510-3_1

proper coordination and implementation. This emerging shift is, in many ways, comparable to the early development of counterterrorism strategies within police organisations in the 1960s and 1970s (Hobson & Pedahzur, 2022), highlighting both the challenges and the urgent need for systematic integration.

My hope for this book is not only to contribute to the steadily emerging academic discussion and empirical evidence on these issues within criminology, and policing studies in particular, but, most importantly, it will reach police practitioners, without whom these debates are meaningless. Alongside biodiversity loss and air pollution, climate change is one of the most significant, society-wide, environmental challenges facing humanity, pushing everyone into new territory of challenges and adaptation efforts (United Nations Department of Economic and Social Affairs, 2024). The anthropogenic impact on climate change, meaning it is caused by human activity, presents a perplexing paradox: it is the very contemporary ways of life, work, and the pursuit of development and profit that humanity has created are driving this catastrophic scenario. Another unfortunate realisation is that any positive developments in this area, such as the growing momentum for ecocide legislation, additional resources for setting up environmental police teams, or city-wide heatwave plans, are driven by the accelerating, increasingly adverse, and more noticeable effects of climate change and environmental degradation. Although the book is aimed at the police as the primary audience, it also speaks to us—and you as an individual reader—emphasising that we all can create bridges of understanding and collaboration in this field.

Emerging scholarship and grey literature on this topic tends to focus on one of two key questions. The first one predominantly explores how the consequences of climate change are transforming the landscape of harms and crimes (Agnew, 2012; Farrall, 2012; Hallenberg, 2021; Matczak & Bergh, 2023; UNODC, 2024). Agnew's theoretical model of the effects of climate change on crime suggests that climate-related consequences of climate change will likely increase a wide range of societal strains that contribute to criminality, heighten social conflicts, reduce both formal and informal social control, weaken social support, foster beliefs and values conducive to crime, and increase overall opportunities for crime (Agnew, 2012). Second question has generated a growing body of scholarly evidence suggesting that climate change, alongside other complex challenges, will not only have a significant impact on policing but may also lead to a long-term transformation in the role of law enforcement within societies. This shift will increasingly require police forces to demonstrate resilience and collaborate with a diverse range of actors who are also engaged in policing-related tasks (Carrilho, 2019; Mutongwizo et al., 2022; Blaustein et al., 2023, Lydon et al., 2024; Malik & Berg, 2024).

Featuring interdisciplinary research, a comprehensive literature review, and 23 expert interviews, this book builds on and extends my earlier co-authored research work on the impact of climate change on police professional practice (see Matczak & Bergh, 2023). It is written with the aim to help police practitioners, policymakers, scholars, and others interested in the topic, better understand the intersection between climate change and police work, as it manifests through different processes, practices, and events, along with its nature and complexity. On a more practical level, this book is intended to assist police leadership in guiding future practice and policy developments in this area and in developing targeted climate adaptation plans within their respective police organisations.

As these are pioneering years, our understanding of the connection between climate change, police, and policing is evolving in real time. To capture a comprehensive perspective, I invited a diverse range of experts from various disciplines to contribute to this research. Among them are criminologists, policing specialists, security experts, climate scientists, police officers, lawyers, and activists. While some agreed to be identified in this book, others preferred to remain anonymous. Their insights and contributions will be introduced gradually as the findings unfold.

The expert interviews included participants who, under ordinary circumstances, would be unlikely to engage in discussions within the same setting. They came not only from different countries but also from distinct organisations and areas of expertise. In total, experts from 11 countries participated, including Australia, Belgium, France, Greece, India, Indonesia, Slovenia, South Africa, Spain, the Netherlands, the United Kingdom, and the United States. Among them were professionals with experience and insights from institutions such as EUROPOL, the United Nations Office on Drugs and Crime (UNODC), the International Criminal Court (ICC), Stop Ecocide International, and the World Bank.

Clifford Shearing, prominent and renowned policing scholar, interviewed as one of the experts, pointed out, we are currently witnessing a significant restructuring of police at its conceptual, organisational, and practical levels—a process that is far from systematic or planned:

> We live in a time of a huge flux in policing. It's like we took the Lego of policing and just threw it up in the air. And we don't know how it's going to land, but certainly it will not be as before. (Clifford Shearing)

In the field of criminology and policing studies, David Lydon, a former police officer who is now a Senior Lecturer in Professional Policing in the United Kingdom, highlighted a significant gap in scholarship. He pointed out a notable paucity of researchers capable of analysing current trends to develop theoretical frameworks for understanding the future trajectory of policing in the context of emerging global challenges.

> Arguably there are at the moment very few people in police scholarship who are what I would call futurists. They're dealing with the here and now, but they're not futurists in thinking in that way. Being a futurist is more of an interdisciplinary approach. (David Lydon)

The multilayered complexity of this issue means that the impact of climate change can no longer be researched or discussed solely from the perspective of a single discipline in order to construct a comprehensive narrative. Recent climate change debates have elevated the interdisciplinary approach in science to new heights. Accordingly, this book draws on literature and expert perspectives from different branches of criminology and beyond, in an effort to contribute to that shift, and to capture and understand at least a small fraction of these 'falling Lego pieces'.

We live in a world of constant transformation. The issue of how climate change affects police work emerges alongside an equally significant and unpredictable technological revolution, accelerated by the advent of AI, which already profoundly influences the crime and harm landscape and police preparedness to face these challenges. Additionally, this book's topic arises amidst ongoing discussions about the impact of social media, societal polarisation, global geopolitical instability, geoengineering experiments, and numerous other interconnected factors. As US investigative journalist Christian Parenti (2011) has long observed that the consequences of climate change intersect with pre-existing crises and vulnerabilities, which he describes as a 'catastrophic convergence', which is a situation where problems compound and amplify one another, with one crisis manifesting through another, also known as a 'cascading failure'. While the consequences of climate change are inevitable, the scale and extent of these changes remain uncertain. Climate change creates a ripple effect, often likened to the butterfly effect, where its impacts cascade unpredictably. Uncovering these ripple effects and their manifestations is challenging, as they rarely come with a direct label, and this holds true for police work as well, as highlighted by Thammy Evans, another interviewed expert who is Senior Research Associate in Climate Change and (In)Security Project based in the United Kingdom:

> Climate issues don't appear as climate issues directly. It's a bit like dealing with poverty. The police don't really deal with poverty, but they do deal with its symptoms.
> These symptoms don't come with a label that says this is a climate induced effect.
> (Thammy Evans)

Although climate change is a gradual process, it often reminds us of its presence through acute and potentially catastrophic events, such as disasters. However, the challenges it induces have gradually permeated the police and policing landscape, both directly and indirectly. This book seeks to connect these dislocations of climate change manifestations and expand the ongoing debate about its implications for police work.

Methodology

Although this book is based on 23 interviews, conducted between April and November 2024, I initially reached out to twice as many potential experts. The interviewees were recruited through a snowball sampling approach, beginning with a few individuals already recognised in the field for their research and publications on this topic. Some experts were specifically approached due to their distinctive knowledge or experience in particular aspects of the discussion, such as ecocide or the relationship between crime and extreme heat.

All interviews were recorded, transcribed, coded, and thematically analysed. While the selected and presented quotes in this book do not encompass all interviews due to the word count limitations, the entirety of the interview data has informed and shaped the book's structure and content. Additionally, the interviews have played a crucial role in analysing and contextualizing the findings in relation to the existing literature on the topic.

As the book aims to provide a broad overview of the issue, it was essential to incorporate global perspectives, particularly from representatives of international organisations that are increasingly engaging in this field. Due to geographical constraints, the majority of interviews were conducted online, with only three conducted in person. While I made a concerted effort to include voices from the Global South, time and language constraints limited the extent to which this could be fully achieved.

References

Agnew, R. (2012). The ordinary acts that contribute to ecocide: A criminological analysis. In N. South & A. Brisman (Eds.), *Routledge international handbook of green criminology* (pp. 52–67). Routledge.

Blaustein, J., Miccelli, M., Hendy, R., & Hutton-Burns, K. (2023). Resilience policing and disaster management during Australia's black summer bushfire crisis. *International Journal of Disaster Risk Reduction, 95.* https://doi.org/10.1016/j.ijdrr.2023.103848

Carrilho, L. (2019). *Police as first responders to the global climate crisis.* https://medium.com/we-the-peoples/police-as-first-responders-to-the-global-climate-crisisd4896f4d8ec. Accessed 14.02.24.

Farrall, S. (2012). Where might WE be headed? Some of the possible consequences of climate change for the criminological research agenda. In S. Farrall, T. Ahmed, & D. French (Eds.), *Criminological and legal consequences of climate change. Onati Institute for the Sociology of law* (pp. 7–26). Hart Publishing.

Hallenberg, K. (2021). Crime, harm, and climate change nexus. In W. Leal Filho, A. M. Azul, L. Brandli, A. Lange Salvia, P. G. Özuyar, & T. Wall (Eds.), *Peace, justice and strong institutions* (pp. 1–12). Springer.

Hobson, B., & Pedahzur, A. R. (2022). The Munich massacre and the proliferation of counterterrorism special operation forces. *Israel Affairs, 28*(4), 625–637.

Lydon, D., Hallenberg, K., & Kapageorgiadou, V. (2024). This is not a drill': Police and partnership preparedness for consequences of the climate crisis. *International Journal of Police Science and Management, 27*(1) 16-30. https://doi.org/10.1177/14613557241248295

Malik, A., & Berg, J. (2024). Polycentric governance, epistocracy and the limits of policing knowledge in preparing for the climate crisis. *Policing and Society: An International Journal of Research and Policy,* 1–15. https://doi.org/10.1080/10439463.2024.2389961

Matczak, A., & Bergh, S. I. (2023). A review of the (potential) implications of climate change for policing practice worldwide. *Policing: A Journal of Policy and Practice, 17,* paad062. https://doi.org/10.1093/police/paad062

Mutongwizo, T., Blaustein, J., & Shearing, C. D. (2022). Resilience policing and climate change: Adaptive responses to hydrological emergencies. *CrimRxiv,* 1–19. https://doi.org/10.21428/cb6ab371.fb7dfcd0.

Parenti, C. (2011). *Tropic of chaos. Climate change and the new geography of violence.* Nation Books.

United Nations Department of Economic and Social Affairs. (2024, July). SDGs in practice: Climate change, biodiversity loss, and pollution. https://sdgs.un.org/sites/default/files/2024-07/SDGs%20in%20Practice%20-%20Climate%20etc.pdf. Accessed 12.12.24.

United Nations Office on Drugs and Crime. (2024). *Global law enforcement in the harm landscapes of climate change.* https://www.unodc.org/documents/data-andanalysis/Crimes%20on%20Environment/UNODC_GLOBAL_LAW_ENFORCEMENT_IN_THE_HARM_LANDSCAPES_OF_CLIMATE_CHANGE.pdf. Accessed 31.03.25.

Chapter 2
Setting the Scene

How Bad Are the Accelerating Effects of Climate Change?

During the peak period of writing, the year 2023 was declared the hottest on record (Bui et al., 2024). Furthermore, the calendar year 2024 was reported as the first to exceed 1.5 °C above pre-industrial levels, marking the initial occurrence of such warming within a 20-year period averaging 1.5 °C (Bevacqua et al., 2025). Why has it deteriorated so rapidly?

Meteorologists and climate scientists define climate as the range of weather patterns observed over an extended period, typically using a 30-year time frame, which includes factors like temperature fluctuations, rainfall amounts, or hours of sunshine (World Meteorological Organisation, 2021). In 2022, the La Niña phenomenon, one of the El Niño-Southern Oscillation (ENSO) climate patterns that break the normal weather conditions, was unusually long and resulted in higher rainfall to parts of Asia and the Pacific, and severe drought and extreme temperatures around the Horn of Africa and parts of the Americas and Europe (IDMC, 2023). The 2021 Intergovernmental Panel on Climate Change (IPCC) report clearly states that both average and extreme heat measurements are rising across all continents. A heatwave that had a 1 in 10 chance of occurring in any given year in the pre-industrial climate is now 2.8 (1.8–3.2) times more likely to happen and will be 1.2 °C hotter. With 2 °C of global warming, the same heatwave would occur 5.6 (3.8–6.0) times more often and would be 2.6 °C hotter (Clarke et al., 2022). By 2015, climate change had raised the likelihood of daily precipitation extremes exceeding the 99.9th percentile of pre-industrial events by an average of 18% (ibid.).

However, unlike temperature changes, these precipitation shifts vary significantly across different regions and seasons (Fischer & Knutti, 2015). AI-assisted analysis predicts that even with substantial reductions in greenhouse gas emissions, global warming is still on track to continuously surpass the 1.5 °C threshold above pre-industrial averages by the early 2030s (Diffenbaugh & Barnes, 2023). A 2-degree rise in global temperatures, referred to as the 'Crossing Year', is projected

© The Author(s) 2025
A. Matczak, *Adapting to Climate Change in Modern Policing*, SpringerBriefs in Criminology, https://doi.org/10.1007/978-3-031-97510-3_2

to occur in the 2040s (if no significant policy changes are implemented), and is considered a critical threshold that could trigger a range of cascading and dangerous consequences for humanity (Park et al., 2022). It is also becoming clear in the scientific community that even with the immediate achievement of net-zero emissions, the carbon already trapped in the atmosphere will continue to influence ecosystems, human populations, and infrastructure (Donatti et al., 2024). A recent and urgent appeal by 15,000 scientists highlights that, despite 6 IPCC reports, 28 COP meetings, countless other reports, and tens of thousands of scientific papers, progress on addressing climate change has been minimal. This is partly due to strong opposition from those who profit financially from the current fossil-fuel-based system, which is a major contributor to global warming (Ripple et al., 2024).

The current patterns of human activity and resource consumption are fundamentally incompatible with long-term ecological sustainability. The primary factors contributing to human-induced impacts on Earth's energy system are the emissions of greenhouse gases and aerosols, along with changes in surface albedo (Richardson et al., 2023). Scholars warn that, due to the rapidly changing climate, data from three decades ago no longer accurately reflect current conditions. Although the IPCC reports have become a default reference point to communicate scientifically about climate change, the process of preparing and negotiating these reports are deeply political and present more optimistic scenarios than it would have otherwise been without the political pressure (see Hughes, 2024). The increasing frequency of extreme weather events suggest that we may not yet have experienced the full extent of the extremes that today's atmospheric and oceanic warming can produce. Even more concerning is that we often fail to adequately investigate and learn from these extreme weather events, which are quickly forgotten (Lee et al., 2024).

> The Anthropocene offers a similarly catastrophic threat landscape as a nuclear winter, but its vision of extinction – slow, latent, barely discerned, and yet violently effective – exists in stark contrast to the spectacular immediacy of nuclear war and deterrence logics. The Anthropocene extinction isn't solved with a telephone hotline. (Harrington & Shearing, 2017, p. 15)

NASA researchers have explored the question, 'What will global land climate look like with 2°C of warming?' They found that, with regional variations, there will be significant changes in six key climate factors: average air temperature, precipitation, relative humidity, incoming shortwave and longwave radiation, and wind speed. These changes are expected to directly increase the risks of heat stress and wildfires (Park et al., 2022). Countries near the equator will face prolonged periods of 'extreme days', with more than a quarter of the global population potentially experiencing an extra month of severe heat stress each year compared to the mid-twentieth century (1950–1979) (ibid.). In addition to extreme heat, intense precipitation events could lead to significant socio-economic damage through cascading disasters like floods and landslides. The projected climate of the 2040s is also expected to impact renewable energy potential, particularly wind, solar, and hydropower (ibid.). The cost of postponing or (in)action was long estimated to be equivalent of losing between 5 and 20% of GDP each year (Stern, 2006).

Although climate change and its consequences are the focus of this book, it is only one of many elements contributing to the largely uncharted planetary territory we now find ourselves in. The United Nations views climate change as part of a broader 'triple planetary crisis', (UNFCC, 2022) along with air pollution and biodiversity loss. Climate change and the biosphere are two essential, highly interdependent components of so-called planetary boundaries framework, which is grounded in Earth system—the planet's interconnected physical, chemical, and biological processes. This framework identifies nine critical processes that are vital for maintaining the stability and resilience of the Earth system as a whole.

Currently, six of these nine processes have been significantly breached, and all are deeply impacted by human activities. Increasing evidence suggests that the current level of boundary transgression has already pushed the Earth system beyond a 'safe' threshold (Richardson et al., 2023).

The June 2021 Pacific Northwest 'Heat Dome' event, where an unforeseen high-pressure system trapped hot air over the region, resulting in record-breaking temperatures, highlights the limitations of our current understanding in this area. The unprecedented intensity and magnitude of this event, where temperatures soared well above 40 °C (104 °F), exceeded most projections for the region, making it more severe than many experts had anticipated (Li et al., 2024). Furthermore, the movement of Earth's magnetic poles (O'Callaghan, 2018) and the occurrence of solar flares (National Oceanic and Atmospheric Administration, n.d.) illustrate that there is still much we do not know how climate change may evolve and be influenced by nonhuman factors. The gravity of the situation is apparent, but the limits of our understanding suggest we may still be underestimating just how treacherous the path ahead truly is.

Climate Change Comes with a Time Lag and Uneven Distributions of 'Hurt'

Climate change is a slowly unfolding and largely 'invisible' phenomenon, and as one of my interviewees, Jojo Mehta, co-founder and current chair of Stop Ecocide, remarked: 'I think it's quite human, as in people don't tend to pay attention to things until they're right in front of them'. The idea that people generally, not only the police, underestimate the impact of climate change was widely discussed by my interviewees. Alexandra Jones, a senior police officer and advisor from the Dutch Police pointed out that people are programmed to associate low frequency events with low probability, which makes it difficult to prepare for and take more decisive actions:

> Climate change has a lag phase. It's gradual and then you get a tipping point, then you get a typical long phase, and you get an acceleration. Then you get a plateau phase and get an acceleration. This is very difficult for us. We tend to manage this as an incident, through late incorporation. When something has a low frequency or it's not immediately visible, we tend

to underestimate the probability. These are separate things. We don't cognitively deal well
with it and it's very difficult to explain. [...] We may well be in an acceleration phase, and
we're still managing this as a series of incidents and unrelated events. (Alexandra Jones)

Although climate change is a global issue that exists everywhere, it needs to be
acknowledged that the magnitude of climate change effects is geographically
unevenly distributed, and its impacts are regionally disproportionate. For example,
the South Pacific region, comprising the island states of Micronesia, Melanesia,
and Polynesia, is considered to be one of the most vulnerable to climate change
(Bueger, 2020). It's important to recognise that countries in the region experience
different challenges, legacies, vulnerabilities, and adaptation strategies. The levels
of country's vulnerabilities are determined by the size of their economies, physical
infrastructure, coastline-to-landmass ratios populations, lingering effects of colo-
nialism, as well as law enforcement capacities. Similarly, the impacts of climate
change on police services vary depending on their geographical location, the extent
and type of climate change impact, socio-political circumstances, and how well-
equipped police services are to confront these challenges. It is therefore crucial to
acknowledge the varying levels of vulnerability across countries and police services in relation to climate change. While the Notre Dame Global Adaptation
Initiative Country Index (ND-GAIN) measures a country's exposure, sensitivity,
and adaptability across six vital sectors (food, water, health, ecological services,
human habitats, and infrastructures) it may also offer insights into how prepared
police services in those countries are to face the adversities brought of climate change.

The consequences of climate change are not only unevenly distributed geographically (horizontally) but also vertically within society and across different sectors.
In the rapidly growing body of climate change-related literature, climate change is
now being discussed as a health crisis, a security crisis, justice matter, and a governance crisis. The writings emphasise that the burden and costs of climate change
will be shared, but the exact distribution of this burden, or 'hurt', as one expert
eloquently put it, remains to be determined:

I don't think we can fix climate change. Every intervention makes the system change and
requires a new adaptation. Politicians, because of the way they are held to account in the
media and by the people themselves, live from one week to the other. They have to offer a
solution. I would like people in general to be more honest. I would like politicians to be
more honest, too, and I would like them to say that we are not going to solve this. We can
only adapt and it's going to hurt because change always hurts. The discussion in politics is
always about distribution and allocation.

How are we going to distribute the hurts? (Alexandra Jones)

Although not always framed in these terms, discussions regarding who will be
affected by climate change, how they will be impacted, and the likely timeline of
these consequences have long been present in security studies, often driven by military and security experts. However, criminologists and policing scholars have either
overlooked or dismissed these discussions, in consequence limiting critical engagement with the issue.

Military-Driven Discourse on Climate Change: Friend or Foe?

Raising the question of the military's role and/or its academic presence in criminology or policing studies is a challenging task, as it inevitably brings up concerns about securitisation policies and practices, much influenced by the excessive militarisation of the police in the United States (see Balko, 2014; Kraska & Kappeler, 1997) but also other parts of the world. However, while researching the intersection of climate change and policing, it is impossible, and even inappropriate, to overlook the scholarship, practice, and advocacy from security studies. This field already has a substantial body of academic and grey literature on the issue, and these increasing contributions have significantly shaped the debate by framing the consequences of climate change as threats and/or risk multipliers (see Metis Institute for Strategy and Foresight et al., 2024; Middendorp & van Campen, 2023; Oels, 2012).

At first glance, the discussions surrounding climate change and the roles of the police and military appear to share common ground and may even seem interchangeable. This is likely why some police services seek advice and guidance from the military first (e.g. as I observed in the Netherlands or in the United Kingdom). However, upon closer inspection, the two discourses may introduce different logics into the conversation with distinct methods implied by the language they use. In the English language, the term 'security' often evokes a range of confrontational practices, rooted in a zero-sum mentality that disregards context and introduces a warlike logic. This mindset can undermine the cooperative efforts necessary to address environmental challenges (Deudney, 1990). Security conveys a sense of urgency, acceptance of exceptional measures, and often an unwillingness to negotiate, sometimes leading to the 'securitisation' of issues where decisions are made unilaterally by a political community. Contemporary security thinking is fundamentally based on sovereign decisions to protect a community, typically through the identification and construction of an enemy (Harrington & Shearing, 2017).

While concerns about the impact of climate change on police practice have only recently emerged from the shadows, the development of thought and scholarship on environmental security (or climate change and security) has a longer tradition, beginning in the 1980s, and heavily involving the military sector as one of the central actors to address the issue (Trombetta, 2008). Climate change began to be considered a security issue by science and politics, particularly between 2007 and 2009 (Oels, 2012). The increasing awareness and recognition of climate change within the military sector has been a subject of ongoing debate in security studies, where the military is progressively framed as part of the solution. Efforts to 'green' defence not only aim to enhance operational sustainability but also serve to reframe the military's image from that of a major polluter and environmental disruptor to a proactive actor in climate mitigation and adaptation (Jayaram, 2021). One of the contributing factors to this development in some parts of the world was the military's direct experience in responding to disasters, as portrayed by a security expert from India:

> They [military] knew this was something that was very strategic in nature and had to be integrated, particularly because the military is involved in very ecologically fragile areas in

India, like the Himalayas. These areas are particularly fragile. We have seen how the recent disasters have increased the burden on the military. Even if you look at the impacts on the military, that is something that cannot be denied and the military acknowledges it a lot more, as compared to what it used to before.
(Interviewee 5)

With the launch of the The North Atlantic Treaty Organisation (NATO) Climate Change and Security Centre of Excellence in 2023 in Montreal (Global Affairs Canada, n.d.), one might interpret this development as an indication that even the military sector is changing and undertaking more of a humanitarian role in response to accelerating effects of climate change, something that was articulated in my interview with Thammy Evans, too.

Countries are using their military to combat illegal wildlife trafficking, counter deforestation or illegal gold mining. You know, this is all starting to get into this space where suddenly militaries are not about greening defense. It's about defending the green.
(Thammy Evans)

However, similar to the police, the nexus between the military and climate security is complex and not easily defined or broadly represented. Dhanasree Jayaram (2021) argues that because different states have varying perspectives and approaches to the relationship between climate change and security, the practical implications of involving the military in climate-related issues are widely debated. Concerns regarding the 'militarisation' of climate change and the potential for 'greenwashing' within this context remain insufficiently addressed. Moreover, according to the author much of the discourse on this topic is dominated by institutions and experts from Organisation for Economic Co-operation and Development (OECD) countries, often lacking the perspectives of other regions around the world (ibid.).

The securitisation of climate change, which frames it as a security threat or 'threat multiplier' aims to heighten its importance and signals that the military can play a role in addressing the issue. However, Dhanasree Jayaram also points out that the notion of 'othering' the climate is evident in the discourse of riskification, which treats climate change as a risk. This perspective, however, emphasises preparation, anticipation, precaution, and prevention, in contrast to the securitisation's focus on defence, deterrence, and combating an external enemy. A key lesson from India is that both the approaches, securitisation and riskification, when advocating for a permanent state of emergency, may actually impede global collective efforts to effectively address climate change (Jayaram, 2021).

From a practical standpoint, military leadership has historically been more engaged in strategic discussions and decision-making processes concerning national security compared to the police. A similar pattern emerges in the context of climate change. For instance, in England and Wales, one of the earliest strategic documents for law enforcement was influenced by military analysis. However, as noted by David Lydon, its content did not fully permeate all levels of police leadership:

The College of Policing produced that report [the Future Operating Environment 2040], which was for internal consumption, really for the UK Police Service. But effectively they just drew from global strategic trends analysis, particularly some of the work that was done by the Americans, the Australians around global challenges. The interesting thing to me was that it seemed to be a very niche document. It quickly became apparent that there were a lot

> of people, particularly even senior police officers, who were not even aware of the report, or if they were aware of it, they hadn't read it and they weren't aware of what the content of it was, what the implications were. This was quite interesting and disturbing at the same time. (David Lydon)

Nonetheless, many of my interviewees, when asked about the role and influence of the military in addressing the impact of climate change on police work, raised several interesting points, many of which could be considered as interesting lessons learned for the police, which I will elaborate and discuss below. The first observation is that the military tend to adopt a system thinking approach, which emphasises the importance of adaptations and trade-offs, an approach the police could benefit from embracing more fully.

> These are all sorts of crises that are intertwined. They're the complex problems in themselves and they are complex mega problems as they become entwined. If there's one thing, I would like the police to learn from the military, is to understand that these are so-called complex systems and that a linear policy won't help. Our politicians typically advocate linear interventions, making us think that the problem is solved. There are no solutions, there are only adaptations and trade-offs. The military are far better aware of this. When you're in a complex system you are part of the problem. (Alexandra Jones)

Another key lesson from the military sector is its strong aversion to 'surprise', which stems from its emphasis on thorough planning, strategic thinking, and scenario-based training. This strategic mindset, characterised by proactive planning, a scenario-driven approach, an outward-looking perspective, and the pursuit of alliances, was strongly highlighted in my interviews with Thammy Evans, Alexandra Jones, and Matt Luna, the founder of The Hague Roundtable on Climate and Security. Adopting such practices could significantly aid the police in shifting from a predominantly reactive institution to one that is more proactive and preventive.

One important observation, and potential valuable insight for the police, about the military was how two specific types of military personnel became the pioneers of climate change discussion within the sector due to their early exposure to its implications. In my conversation with Thammy Evans it was said that military personnel tend to be more engaged with environmental issues and crimes, partly due to their extensive time spent outdoors, which fosters a natural affinity for environmental concerns. Additionally, according to Thammy, two specific groups within the military demonstrate a particular inclination towards environmental considerations. The first is the logistics corps, or supply personnel, who quickly recognise the operational benefits of sustainability in the United Kingdom, and who are logisticians that have been at the forefront of 'greening' defence initiatives. Logistics, or logistical chain, is always the biggest cost for any military operation. Greening logistics, and making them operationally independent and autonomous, is a significant added value here. Reducing supply needs for frontline operations enhances efficiency, making sustainability a strategic advantage. The second pioneering group in the United Kingdom according to Thammy is military engineers, who naturally approach problems through a systemic lens. Their ability to identify root causes and interconnections allows them to grasp environmental challenges more readily. If this observation is true in other sectors and among other professions, the police leadership should identify their respective 'logisticians' and 'engineers' within the police service as they might be the ones with the highest potential to understand the implications of climate change on their work.

The relationship between criminology and security studies remains uneasy, with growing reluctance within criminology to regard the military as a trustworthy stakeholder (see McGarry, 2024). However, adopting a whole-of-society approach, a much celebrated academic framework that addresses complex challenges through the coordinated involvement of all relevant sectors, stakeholders, and communities is essential for tackling climate change. The police, on the other hand, are consistently engaged with the public. The whole-of-society or systemic approach is further conceptualised by Blaustein et al. (2024) as a mechanism for promoting adaptive governance in security management during the Anthropocene, which is an era characterised by the dissolution of traditional boundaries between humans and nature. This approach requires fostering collaboration and exchanging insights whenever possible. Such engagement is particularly crucial given that the military's role in responding to climate-related events, such as natural disasters, is typically intermittent rather than a default component of such operations.

The Risk of Leaving the Police behind

Within criminology, environmental matters have long been addressed and researched from the perspective of its quickly developing sub-branch, green criminology. Green criminology focuses on harms impacting both human and nonhuman species, the environment, and the planet. Its ultimate aim is to inspire change and achieve environmental and ecological justice (Lynch, 1990; Beirne & South, 2007; White, 2008). In order for criminology to remain relevant is to engage more with environmental matters, consider topics such as mass extinction and/or planetary protection (White & Hasler, 2019). Unfortunately, despite the growth of green criminology, and with few exceptions, the reference to the police is quite often absent. As environmental problems worsen, the role of public safety sectors, particularly the police, is critical in the pursuit of environmental sustainability, leaving the police with no other choice but to evolve (Konyk, 2018). This could be due to green criminology being closely coupled with critical criminology, as indicated by a critical and 'green' criminologist based in the United Kingdom:

> It's become taken for granted that corporate crimes and white-collar crimes are more extensive, create more harms, create more victims than all other forms of crime. I think that's kind of accepted in the discipline of criminology. I suppose that's something that's changed for the better. But in a way it hasn't changed for the better because although that's recognised, there are still not very many corporate crimes and white-collar crime scholars. I think that's for a good reason. I think that being involved in studying and writing about corporate crime and white-collar crime within criminology, what you're going to be doing is critiquing criminology, finding a way to critique the criminal justice system. It's an agenda which is never going anywhere. It's only going to end up as a critique. (Interviewee 15)

Apart from the growth of green criminology without a corresponding development in 'green policing', substantial scholarship on plural policing has steadily documented not only the increasing pluralisation of policing tasks, shared with other actors, and the growth of the private security sector as a major competitor, but also the gradual decline of law enforcement as the primary state policing actor (Crawford, 2006; Johnston, 1992; Jones & Newburn, 2006; Shearing & Wood, 2003). The concept of plural policing purports that although policing remains directly provided and overseen by national and local government institutions, it has expanded beyond this traditional framework to include various other forms of policing. These include private security services contracted by the government, transnational policing arrangements operating at a supra-governmental level, commercial markets for policing and security that function independently of the state, and community-led policing initiatives at the grassroots level. This evolving landscape reflects a world where policing is increasingly pluralistic and interconnected through complex networks (Loader, 2000).

The debate around defunding the police in the United States (Lum & Nagin, 2021) and United Kingdom (Fleetwood & Lea, 2022) has further questioned the role of state police.

Additionally, the varied approaches to policing climate and environmental demonstrations have also contributed to negative perceptions regarding the legitimacy and role of the police in addressing modern challenges (Di Ronco, 2023). In the introduction to their article on the role of the police during 2019–2020 Australia's Black Summer bushfires, Blaustein et al. (2024) note that policing has also traditionally been peripheral to both disaster resilience research and emergency management. All of this has created a situation where the police risk being excluded from almost any climate change-related discussions, potentially leaving them behind in these conversations and positioning them as adversaries.

The police as a security actor within a broader governance constellation have almost always been defined in anthropocentric terms. According to Harrington and Shearing (2017) modern security approaches have traditionally viewed the Earth as a stable system functioning independently of human influence, and the concept of an 'Earth system' has been largely absent from most security frameworks. Therefore, the climate change crisis necessitates a fundamental reshaping of not only security studies but also criminology (Holley et al., 2020). The Anthropocene is pushing the boundaries of criminology scholarship by disrupting a differentiation between humans and nature, a distinction that bred previous conceptualisations and framings of harms, security, and policing. In the Anthropocene, the occurrence of disruptive events that break normalcy and introduce exceptional conditions is becoming more frequent, often with little recognition or acknowledgement. This unquestioned belief in the security provided by nature, regardless of human actions, is now being disrupted and requires new lens of analysis (Harrington & Shearing, 2017).

The impact of climate change on police work necessitates a new analytical framework. Within this discourse, which is primarily shaped by military and disaster studies, criminology and policing studies remain peripheral. This marginalisation is further reinforced by the long-standing critique of police within criminology.

However, as the following chapters will demonstrate, law enforcement is, and will increasingly be, among the institutions most profoundly affected by climate change. A lack of preparedness could have severe consequences, including chronic underfunding, inadequacies in police training and education, challenges in recruitment, and declining public confidence and trust in the police.

Climatisation of Police and Policing

Previous research has found that presenting climate change through different frames, such as a health issue, a national security risk, or an environmental problem, can influence audiences in varying ways (Myers et al., 2012). In this book, and in an effort to avoid othering climate change as merely a risk or threat, I have followed the scholarly work of Angeal Oels and Dhanasree Jayaram and decided to apply the climatisation framework as a way to understand how the consequences of climate change permeate police work, the police organisation, and what is needed at the individual level for police officers to respond more efficiently to these issues. In short, the accelerating impact of climate change does imply new roles for the police and other policing actors, along with different, or modified, means for providing safety and security. The climatisation framework elucidates how existing policing practices are not only adapted to address climate change but also how new practices emerging from climate policy are integrated into the field of policing. These two simultaneous processes collectively shape the impact of climate change on police and policing, as referenced in the title of this book.

Angela Oels (2012) observed that climate change began to be discussed in alarmist terms as early as the 1980s, and that its articulation as a security threat has been present in scholarly literature and political declarations since 2003. Drawing on the Paris School of security studies, she proposes shifting the focus away from looking at the consequences of climate change through exceptional measures towards the analysis of routine practices, and conceptualising climate change as a human security issue. The climatisation framework serves as an analytical lens to examine how existing security practices, including those of the police, military, and intelligence agencies, are mobilised to respond to the impacts of climate change, and how climate policy tools such as climate modelling, early warning systems, and scenario planning are integrated into the security field. Although the police are implicated in these processes, Oels' analyses have primarily focused on the military, migration, and development sectors, thereby further underscoring the marginalised position of the police in these discussions.

The framework of climatisation was later used by Jayaram (2021) to analyse these processes within the Indian military, revealing the significant impacts of climate change on the military as well as its contribution to the national climate goals. The choice of the climatisation framework was dictated in her work by the need to upturn the dominant language of securitisation. Jayaram's work shows how the military has gradually moved towards mainstreaming climate change, categorising these climatising efforts as symbolic, strategic, precautionary, and transformative. None of these are static, something that can be categorised as symbolic might turn transformative when the right conditions allow. This classification will provide a useful guidance in interpreting the content of this book.

Building on these foundational works, I also try to use a more neutral language, than the one of securitisation or riskification. This book extends the climatisation framework by approaching the police from a criminological perspective, conceptualising them both as security, environmental, and criminal justice actors. Policing is understood here as a socially constructed and dynamic process, the boundaries of which are continuously negotiated and shared with a range of other actors. While the concept of plural policing will be explored later in the book, the primary focus remains on state policing and policing by the police. Applying the climatisation framework to this context implies that existing police practices are increasingly required to respond (knowingly or unknowingly) to the multifaceted challenges posed by climate change, while practices derived from climate policy, such as sustainability initiatives, are gradually being integrated into the domain of policing. Although climatisation may involve both mitigation and adaptation measures, the worsening climate conditions outlined earlier in the chapter indicate that police efforts will likely concentrate predominantly on adaptation.

As pointed out by Alexandra Jones, the climatisation of police and policing can also be understood through the interaction between shocks and stressors—events and phenomena whose analysis might assist communities and systems build resilience. Shocks are external short-term deviations from long-term trends, such as disasters (Zseleczky & Yosef, 2014, cited in Frankenberger et al., 2015). Shocks can have a gradual or sudden onset, be transitory, intensive or extensive, seasonal or structural, and vary in frequency. Drawing on Roché and Fleming (2025) recent analysis of shocks within policing studies, climate change-related shocks can be conceptualised as exogenous in nature, that is, originating from broader societal disruptions such as economic crises, political instability, or pandemics. The other type of shocks discussed by the authors are endogenous one, which are internal in nature, stemming from within policing institutions themselves, such as through misconduct or the use of contentious tactics.

While not all climate change-related shocks have become more frequent, many are increasing in their severity, scope, and impact. For instance, hurricanes such as Hurricane Harvey (2017) and Hurricane Ian (2022) were not necessarily more frequent than past storms, but their intensity, rainfall, and resultant flooding were significantly greater, leading to widespread destruction and displacement (National Oceanic and Atmospheric Administration, 2017; National Hurricane Centre, 2022). Stressors, on the other hand, are long-term pressures (e.g. environmental degradation, overpopulation, political instability) that undermine the stability of a system and amplify vulnerabilities (Bujones et al., 2013 cited in Frankenberger et al., 2015). Shocks and stressors may be interrelated and/or occur simultaneously, making it essential to understand the dynamics between the two to analyse the viability of resilience, and the ability of communities/systems to recover from them

(Frankenberger et al., 2015). Therefore, climatisation of the police and policing can also be analysed as a response to the interplay between external shocks (e.g. climate-related disasters) and persistent stressors (e.g. environmental or political pressures), both of which challenge institutional resilience and demand adaptive policing strategies.

The most transformative aspects of climatisation can first be observed in the evolving and changing landscape of crimes and harms, which shifts the attention away from traditional and individual profiles of offenders towards other actors such as states and corporations, as well as collective experiences of victimhood. For at least a decade, climate change has emerged as a criminological question, and interest in its implications within the field of criminology has gradually increased (Kramer, 2020). The impact of climate change transforms the nature of crimes and harms, with old crimes manifesting in new forms and contexts (e.g. water theft), and new crimes are emerging out of rapidly changing circumstances and conditions (e.g. carbon emissions fraud, F-gas trafficking, disaster fraud) (White, 2022).

Lydon et al. (2024) argue that this evolving crime and harm landscape and its intersection with policing, can be examined from two perspectives. The first focuses on direct and indirect actions contributing to climate change, while the second examines the unethical, harmful, and potentially criminogenic consequences of climate change. Consequently, climate change debates have emphasised the need to scrutinise crimes of the powerful, such as state-corporate crimes, now often referred to as 'climate criminals' or 'carbon criminals', whose operations inflict most harms to the environment and climate (Goulka et al., 2023; Kramer, 2020). This shift is accompanied by a quickly expanding body of environmental and ecological legislation (e.g. EU Green Deal, ecocide laws, climate litigation precedence) that requires increased (law) enforcement and oversight.

The precautionary and symbolic aspects of police climatisation are evident at the organisational level, for example, through the incorporation of sustainability language into police operations, the popularisation of resilience as a core concept, police involvement in emergency and disaster management, and their response to eco-movements. There is evidence that the sustainability language adopted by some police services and partnerships[1] has been influenced and informed by the United Nations' Sustainable Development Goals (UN SDGs), particularly Goal 13, which calls for strengthened actions to combat the impacts of climate change (United Nations Department of Economic and Social Affairs, Statistics Division, 2016). Similar evidence was revealed during my interview with Gorazd Meško, a Slovenian criminologist:

> Last autumn, I organised a national conference about the UN Sustainable Goals in relation to criminal justice and security perspectives. We learned that mayors in the majority of local communities here [in Slovenia] take recommendations of the United Nations seriously. They follow this idea, they think globally, but they try to solve the problems locally. (Gorazd Meško)

---

[1] The case of the Netherlands will be discussed in Chap. 2.

However, the emerging empirical scholarly literature on the topic of police organisational preparedness reports on the mostly slow and inadequate police responses to climate change (see Lydon et al., 2024), which sadly confirms that the strategic aspect of police climatisation is poorly executed. The police involvement in responding to disasters, and the role of the police and their partners in planning for and dealing with the resulting civil emergencies, is considered by Lydon et al. (2024) as second important touchpoint on the climate change and policing nexus.

What remains largely absent in these discussions is an understanding of how climatisation processes affect individual practitioners, how the climatisation of their social environment and workplace translates into their personal mindset, and what is required of them to enhance their organisation's ability to better respond to the consequences of climate change.

The climatisation footprint in police and policing can also have malign consequences. It is also important to note that police, along with other criminal justice actors, can be used to counter progressive climatisation processes through mechanisms such as (1) political influence exerted through campaign donations and lobbying, (2) corporate donations to police services to provide additional security at protest locations, and (3) the pursuit of legislation that proactively criminalises environmental protests, as highlighted and elaborated on by Goulka et al. (2023).

Although noticeable and appreciated, the process of adjusting to climate change within the policing realm is a painfully slow, uneven, and politicised process, shaped by level of societal awareness about climate change as well as national and local police conditions. The impact of climate change not always requires 'frontline' reactions, much of it affects the police organisation from within and the purpose of this book is to systematically dissect the avenues through which the police organisations already are and will be increasingly affected by climate change.

References

Balko, R. (2014). *Rise of the warrior cop*. Public Affairs.

Beirne, P. & South, N. (2007). *Issues in green criminology: Confronting harms against environments, humanity and other animals*. Willan Publishing.

Bevacqua, E., Schleussner, C. F., & Zscheischler, J. (2025). A year above 1.5 °C signals that earth is most probably within the 20-year period that will reach the Paris agreement limit. *Nature Climate Change*. https://doi.org/10.1038/s41558-025-02246-9

Blaustein, J., Shearing, C., & Miccelli, M. (2024). Adaptive policing for a climate crisis. *Policing and Society: An International Journal of Research and Policy*, 1–13. https://doi.org/10.1080/10439463.2024.2362713

Bueger, Ch. (2020). *Rising Waters. The Impact of Climate Change on Maritime Law Enforcement Agencies in Pacific Island Countries. Report for the Global Maritime Crime Programme of UNODC*. https://doi.org/10.13140/RG.2.2.25866.44489 Accessed 26.03.24.

Bui, H., Gillett, Z., Neogi, S., Perkins-Kirkpatrick, S., Kim, Y., Alinejadtabrizi, T., Ukkola, A., Aiken, C., Patel, R., Holbrook, N., Richardson, D., Sengupta, A., King, A., Falster, G., Barnes, M., Reid, K., Poncet, L., Bowden, A., Raupach, T., Earl, N., Johnston, D., Rowe, J., Bhagtani, D., Vincent, C., Huneke, W., Udy, D., Meyer, A., Page, J., Kaplish, A., Wilson, A., Arblaster, J., Alexander, L., & Pitman, A. (2024). *The State of Weather and Climate Extremes 2023*.

Australian Research Council (ARC) Centre of Excellence for Climate Extremes, UNSW, https://doi.org/10.26190/92kr-0w80 Accessed 26.03.2024.

Bujones A., Jaskiewicz, K., Linakis, L., & McGirr, M. (2013). *A Framework for Analyzing Resilience In Fragile and Conflict-Affected Situations*. USAID Final Report. Columbia University SIPA.

Clarke, B., Otto, F., Stuart-Smith, R., & Harrington, L. (2022). Extreme weather impacts of climate change: An attribution perspective. *Environmental Research: Climate, 1*(1), Article 012001. https://doi.org/10.1088/2752-5295/ac6e7d

Crawford, A. (2006). Policing and security as 'Club goods': The new enclosures? In J. Wood & B. Dupont (Eds.), *Democracy, society and the governance of security* (pp. 111–38). Cambridge University Press.

Deudney, D. (1990). The case against linking environmental degradation and national security. *Millennium: Journal of International Studies, 19*(3), 461–476.

Di Ronco, A. (2023). *Policing environmental protest: Power and resistance in pandemic times*. Bristol University Press.

Diffenbaugh, N. S., & Barnes, E. A. (2023). Data-driven predictions of the time remaining until critical global warming thresholds are reached. *Proceedings of the National Academy of Sciences, 120*(6), e2207183120. https://doi.org/10.1073/pnas.2207183120

Donatti, C. I., Nicholas, K., Fedele, G., Delforge, D., Speybroeck, N., Moraga, P., Blatter, J., Below, R., & Zvoleff, A. (2024). Global hotspots of climate-related disasters. *International Journal of Disaster Risk Reduction, 108*, 104488. https://doi.org/10.1016/j.ijdrr.2024.104488

Fischer, E. M., & Knutti, R. (2015). Anthropogenic contribution to global occurrence of heavy precipitation and high-temperature extremes. *Nature Climate Change, 5*, 560–564.

Fleetwood, J., & Lea, J. (2022). Defunding the police in the UK: Critical questions and practical suggestions. *The Howard Journal of Crime and Justice, 61*(2), 167–184.

Frankenberger, T., Choularton, R., Kurtz, J., & Nelson, S. (2015). *Measuring Shocks and Stressors as Part of Resilience Measurement*. (Resilience Measurement Technical Working Group Technical Series; No. 5). Food Security Information Network. Retrieved from https://lucris.lub.lu.se/ws/portalfiles/portal/129865219/FSIN_TechnicalSeries_5.pdf. Accessed 15.08.24.

Global Affairs Canada. (n.d.). *NATO Climate Change and Security Centre of Excellence*. Retrieved from https://www.international.gc.ca/world-monde/international_relationsrelations_internationales/nato-otan/centre-excellence.aspx?lang=eng. Accessed 10.09.24.

Goulka, J., Kang, S., Martin, A., Ensign, K., & Beletsky, L. (2023). When crises collide: Mapping the intersections of climate, pollution, crime and punishment. Northeastern University School of Law Research Paper No. 437, Northeastern University Law Review,15(2). Retrieved from https://ssrn.com/abstract=4453693 Accessed 15.02.2025.

Harrington, C., & Shearing, C. (2017). *Security in the Anthropocene. Reflections on safety and care*. Bielefeld: Transcript.

Holley, C., Mutongwizo, T., & Shearing, C. (2020). Conceptualising policing and security: New Harmscapes, the Anthropocene, and technology. *Annual Review of Criminology, 3*, 341358.

Hughes, H. (2024). *The IPCC and the politics of writing climate change*. Cambridge University Press.

IDMC. (2023). *Internal displacement and food insecurity*. Norwegian Refugee Council. https://api.internaldisplacement.org/sites/default/files/publications/documents/IDMC_GRID_2023_Global_Report_on_Internal_Displacement_LR.pdf. Accessed 4.03.2024.

Jayaram, D. (2021). 'Climatising' military strategy? A case study of the Indian armed forces. *International Politics, 58*, 619–639.

Johnston, L. (1992). *The rebirth of private policing*. Routledge.

Jones, T., & Newburn, T. (2006). *Plural policing: A comparative perspective*. Routledge.

Konyk, J. (2018). *Green Policing: Recommended Actions for an Environmental Sustainability Plan for the Vancouver Police Department*. Retrieved from https://sustain.ubc.ca/sites/default/files/2018-54%20Green%20Policing%20Recommended%20Actions%20for%20an%20

Environmental%20Sustainability%20Plan%20at%20the%20VPD_Konyk.pdf   Accessed 16.08.24.

Kramer, R. C. (2020). *Carbon criminals, climate crimes.* Rutgers University Press.

Kraska, P. B., & Kappeler, V. E. (1997). Militarizing American police: The rise and normalization of paramilitary units. *Social Problems, 44*(1), 1–18.

Lee, S. H., Fowler, H. J., & Davies, P. (2024). The climate is changing so fast that we haven't seen how bad extreme weather could get. *The Conversation.* Retrieved from: https://theconversation.com/the-climate-is-changing-so-fast-that-we-havent-seen-how-badextreme-weather-could-get-235726. Accessed 4.10.24.

Li, X., Mann, M. E., Wehner, M. F., Rahmstorf, S., Petri, S., Christiansen, S., & Carrillo, J. (2024). Role of atmospheric resonance and land–atmosphere feedbacks as a precursor to the June 2021 Pacific northwest heat dome event. *Proceedings of the National Academy of Sciences, 121*(4), e2315330121. https://doi.org/10.1073/pnas.2315330121

Loader, I. (2000). Plural policing and democratic governance. *Social & Legal Studies, 9*(3), 323–345.

Lum, C., & Nagin, D. S. (2021). Can we really defund the police? A critical examination of police reform. *Police Quarterly, 24*(4), 350–374.

Lydon, D., Hallenberg, K., & Kapageorgiadou, V. (2024). This is not a drill': Police and partnership preparedness for consequences of the climate crisis. *International Journal of Police Science and Management.* https://doi.org/10.1177/14613557241248295

Lynch, M. J. (1990). The greening of criminology: A perspective for the 1990s. *The Critical Criminologist, 2*(3), 3–12.

McGarry, R. (2024). A criminological–military enterprise. *The British Journal of Criminology, 65*(3), 21–540.

Metis Institute for Strategy and Foresight, Adelphi research, Potsdam Institute for Climate Impact Research, & Bundesnachrichtendienst. (2024). *National interdisciplinary climate risk assessment.* University of the Bundeswehr Munich. Retrieved from https://metis.unibw.de/en/nike/National_Interdisciplinary_Climate_Risk_Assessment.pdf. Accessed 15.02.25.

Middendorp, T., & van Campen, A. (2023). *The climate general: Stepping up the fight.* Éditions La Butineuse.

Myers, T. A., Nisbet, M. C., Maibach, E. W., & Leiserowitz, A. A. (2012). A public health frame arouses hopeful emotions about climate change. *Climatic Change, 113*(3–4), 1105–1112.

National Hurricane Centre. (2022). *Tropical cyclone report: Hurricane Ian (AL092022).* National Hurricane Centre. Retrieved from https://www.nhc.noaa.gov/data/tcr/AL092022_Ian.pdf Accessed 15.02.25.

National Oceanic and Atmospheric Administration. (2017). *Reviewing hurricane Harvey's catastrophic rain and flooding.* Climate.gov. Retrieved from https://www.climate.gov/newsfeatures/event-tracker/reviewing-hurricane-harveys-catastrophic-rain-and-flooding Accessed 15.02.25.

National Oceanic and Atmospheric Administration. (n.d.). *Space weather impacts on climate.* Space Weather Prediction Center. Retrieved from https://www.swpc.noaa.gov/impacts/space-weatherimpacts-climate Accessed 15.02.2025.

O'Callaghan, J. (2018, December 7). Earth's magnetic poles could start to flip. What happens then? *Horizon Magazine.* Retrieved from https://projects.research-and-innovation.ec.europa.eu/en/horizonmagazine/earths-magnetic-poles-could-start-flip-what-happens-then   Accessed 15.02.2025.

Oels, A. (2012). From 'securitization' of climate change to 'climatization' of the security field: Comparing three theoretical perspectives. In J. Scheffran, M. Brzoska, H. G. Brauch, P. M. Link, & J. Schilling (Eds.), *Climate change, human security and violent conflict* (pp. 185–205). Springer.

Park, T., Hashimoto, H., Wang, W., Trasher, B., Michaelis, A. R., Lee, T., Brosnan, I. G., & Nemani, R. R. (2022). What does global land climate look like at 2 °C warming? *Earth's Future, 11*(5). https://doi.org/10.1029/2022EF003330

Richardson, K., Steffen, W., Lucht, W., Bendtsen, J., Cornell, S. E., Donges, J. F., & Rockström, J. (2023). Earth beyond six of nine planetary boundaries. *Science Advances, 9*(37), eadh2458. https://doi.org/10.1126/sciadv.adh2458

Ripple, W. J., Wolf, C., Gregg, J. W., Rockström, J., Mann, M. E., Oreskes, N., Lenton, T. M., Rahmstorf, S., Newsome, T. M., Xu, C., Svenning, J.-C., Pereira, C. C., Law, B. E., & Crowther, T. W. (2024). The 2024 state of the climate report: Perilous times on planet earth. *BioScience, 74*(12), 812–824.

Roché, S., & Fleming, J. (2025). The underplayed importance of shocks in policing studies. *Policing and Society: An International Journal of Research and Policy, 35*(4), 381–397.

Shearing, C. D., & Wood, J. (2003). Nodal governance, democracy, and the new "Denizens": Challenging the Westphalian ideal. *Journal of Law and Society, 30*(3), 400–419.

Stern, N. (2006). *The economics of climate change*. Cambridge University Press. Retrieved from   https://webarchive.nationalarchives.gov.uk/ukgwa/20100407172811/   https://www.hmtreasury.gov.uk/stern_review_report.htm. Accessed 2.03.24.

Trombetta, M. J. (2008). Environmental security and climate change: Analysing the discourse. *Cambridge Review of International Affairs, 21*(4), 585–602.

United Nations Department of Economic and Social Affairs, Statistics Division. (2016, July). *The Sustainable Development Goals report 2016*. United Nations. https://unstats.un.org/sdgs/report/2016/the%20sustainable%20development%20goals%20report%202016.pdf. Accessed 14.03.2024.

United Nations Framework Convention on Climate Change. (2022, April 13). *What is the triple planetary crisis?* UNFCCC. Retrieved from https://unfccc.int/news/what-is-the-triple-planetary-crisis. Accessed 11.06.24.

White, R. (2008). Crimes against nature: Environmental criminology and ecological justice (1st ed.). Willan.

White, R. (2022). Environmental crime and the harm prevention criminalist. *Frontiers in Conservation Science, 3*. https://doi.org/10.3389/fcosc.2022.1049160

White, R., & Hasler, O. (2019). Ecocide: Violence Against the Planet. In W. De Keseredy, C. Rennison, & A. HallSanchez (Eds.), *The Routledge international handbook of violence studies* (pp. 309–319). Routledge.

WMO. (2021). *Weather-related disasters increase over past 50 years, causing more damage but fewer deaths*. World Meteorological Organisation. Retrieved from https://public.wmo.int/en/media/press-release/weather-related-disasters-increase-over-past-50-years-causing-more-damage-fewer. Accessed 10.07.24.

Zseleczky, L. & Yosef, S. (2014). *Are Shocks Really Increasing? A selective review of the global frequency, severity, scope and impact of five types of shocks*. 2020 Conference Paper 5. IFPRI.

Chapter 3
The Changing Landscape of Crimes and Harms

The criminological foundation for understanding crime has always been in constant flux, shaped by processes of criminalisation and decriminalisation and influenced by the social, political, economic, cultural, technological, and historical contexts. However, climate change, unlike any other phenomenon, may inconspicuously transform this field, turning today's ordinary and business-as-usual activities into tomorrow's crimes.

The observable consequences of climate change are generating new risks, expanding opportunities for crime. Although these changes are unfolding gradually, the existing rich body of scholarly, grey literature and practice approaches to environmental crimes offer a valuable foundation for training and preparedness in anticipation of emerging climate-change specific challenges. While not all environmental crimes are directly related to climate change, all crimes associated with climate change fall within one or more of the broader categories of environmental crimes and harms. Certain well-documented environmental harms, such as illegal deforestation or pollution, are for instance intrinsically linked to climate change. While conceptualising the 'climate change-crime nexus', Rob White (2018a, 2018b) distinguishes the following categories: (1) environmental crimes contributing to climate change, (2) environmental crimes occurring as consequences of climate change, (3) public order offences associated with climate change, (4) regulatory offences resulting from policy responses to climate change. This typology was applied and further elaborated on by Hallenberg (2021), which served as a helpful guidance in this book.

For a long time, the omnipresence of hazards in ever-expanding risk societies (Beck, 1992; Mythen, 2014), has made the police, along with other policing actors, increasingly responsible for addressing anything emerging as risks. The criminological and policing literature has recognised these as 'harmscapes' (Berg & Shearing, 2018). However, Harrington and Shearing (2017) point out that we are at a stage where the dynamic nature of the Anthropocene can make any risk or harm

A. Matczak, *Adapting to Climate Change in Modern Policing*, SpringerBriefs in Criminology, https://doi.org/10.1007/978-3-031-97510-3_3

possible, anywhere and anytime, pushing the boundaries of criminological imagination about what crime is and can be in the future. The effects of the Anthropocene are, and will continue to be, unevenly distributed, never completely confined to one location or a specific point in time. This suggests a transition to a world where the distinctions between 'known unknowns' and 'unknown unknowns' are becoming more blurred and challenging to navigate (Blaustein et al., 2024). This uncharted 'harmscape' territory was again emphasised during my interview with Clifford Shearing:

> The earth we exist on today is a new Earth. It has not existed in this form before for humans. I learned this when I was working with insurance companies because they couldn't predict anymore any of these harms. Because all the actuarial tables, historically predictive tables, applied to yesterday's Earth. The idea of once in a hundred-year flood, or once in ten-year flood, or once in a thousand-year flood, didn't make any sense anymore. We have a new Earth, and it is behaving very differently. (Clifford Shearing)

Climate change is transforming the biophysical world, and there is a growing amount of evidence showing how this transformation is radically and rapidly reshaping social, legal, and ecological futures (White, 2015; Kramer, 2020). An interesting example of the emergence of new crimes, harms, or violations, is the case of Slovenia, home to the largest number of beekeepers per capita in Europe, which is part of the country's cultural identity. However, over the recent years, the extreme weather conditions have been causing damage to cross-pollinate fruit trees, which affected honeybee colonies. This led Slovenia to be an active stakeholder and initiator at the international level in the debate over sustainable beekeeping and pollination, banning in 2011 as the first EU country the use of neonicotinoid insecticides.[1] Another example is the 2024 Hajj heatwave and the criminalisation of travel agencies that had illegally facilitated the travel of unauthorised pilgrims. In 2024, the Hajj pilgrimage to Mecca was marked by an extreme heatwave, with temperatures exceeding 50 °C (122 °F). This resulted in the tragic deaths of over 1300 pilgrims, primarily due to heatstroke and dehydration. The intense heat also caused widespread health issues, with thousands more suffering from heat-related illnesses. The heatwave highlighted the growing impact of climate change on large gatherings like Hajj and raised concerns about future pilgrimages, as rising temperatures become a serious risk to participants.[2] This is just the beginning how, in the context of climate change, one can observe that common, acceptable, and routinely practised activities generate a significant negative impact on both the environment and people (Agnew, 2012 cited in Hallenberg, 2021).

---

[1] https://www.politico.eu/article/bees-europe-climate-change-slovenia-agriculture-beekeepers/ Accessed 29.03.2024, https://www.fao.org/newsroom/detail/first-international-forum-for-action-on-sustainablebeekeeping-and-pollination-gives-new-impetus-to-international-cooperation-on-pollinator-protection/en Accessed 3.08.2024.

[2] https://www.washingtonpost.com/world/2024/06/23/hajj-heat-saudi-arabia-pilgrims/ Accessed 1.08.2024.

The Legacy of Environmental Crimes

The first writings on how climate change may impact crime were published over the last 15 years (Agnew, 2012; Farrall et al., 2012). However, the analysis of the climate change-crime nexus must begin by acknowledging the extensive and much longer legacy of scholarship, legislation, and policies in the field of environmental crimes.

The gradual recognition of environmental harms and the subsequent growth of environmental legislation illustrate the evolving crime and harm landscape. Some of the changes include a criminalisation of new types of violations and harms, behaviours that were not previously considered harmful, and some envisage a higher degree of criminalisation. Environmental crimes are now one of the most profitable and estimated to be between the third and the fourth largest criminal sector worldwide, with their monetary value in 2016 estimated between USD 91 and 258 billion annually (Nellemann et al., 2016). The anthropogenic causes and consequences of climate change will undoubtedly deepen and broaden the complexity of these crimes and their impact.

In the joint United Nations Environment Programme (UNEP) and Interpol report (Nellemann et al., 2016:7) environmental crimes are defined as crimes that '*are most commonly understood as a collective term to describe various forms of illegal activities harming the environment and aimed at benefiting individuals, groups, and/or companies from the exploitation of, damage to, trade or theft of natural resources*'. To enhance the analysis, Rob White (2015) proposes categorising environmental harms into (1) crimes/harms of water pollution, (2) crimes/harms of air pollution, (3) crimes/harms of deforestation and spoiling of the land, and (4) crimes/harms against animals/nonhuman species.

The complexity of these crimes necessitates a reimagining of the 'crime scene' as one that is interconnected with organised crime, corporate crime, white-collar crime, and/or safety violations. Historically, environmental crimes have played a secondary role in these intersections and have often been overlooked in investigations, as highlighted in an interview with a senior environmental expert from EUROPOL. However, some specialised environmental crime police units, such as CESANE in France, have begun discussions on shifting the narrative surrounding environmental crimes and assessing the environmental impact of all criminal activities. This evolving crime scene is characterised by extensive and intricate traceability of evidence, much of which exists in cyberspace, making the assessment of its overall impact particularly challenging. Furthermore, the concept of victimhood extends beyond individuals to encompass collective entities, such as communities and ecosystems—for instance, the case of Mar Menor in Spain.[3]

---

[3] These are some preliminary findings from a different project entitled "Approaches to environmental crime enforcement in Europe—Comparative analysis and lessons for the Dutch Police" funded by the Dutch Police Academy.

The lack of a shared consensus on what constitutes an environmental crime represents a major obstacle in combating such offenses. This is strongly articulated and meticulously documented in the most recent 2024 United Nations Office on Drugs and Crime (UNODC) report 'Global Analysis on Crimes that Affect the Environment', where it is stated that the legal protection of the environment is characterized by a fragmented and complex network of international and regional agreements, which are ratified and implemented inconsistently across national legal systems. This ambiguity contributes significantly to discrepancies among national legislations, hindering the harmonization of enforcement strategies, data collection, and the establishment of common standards. This also creates opportunities for both criminal and economic actors to exploit legislative loopholes and enforcement gaps. Without mutual clarity among the sectors and stakeholders involved, consolidating information becomes particularly challenging. While certain offenses, such as poaching, are widely recognized as environmental crimes across the EU, others, such as waste and timber trafficking, are approached with varying levels of priority depending on the country. Furthermore, in some European judicial systems, the absence of a distinct 'environmental crime' category means that offenses like air or water pollution are only investigated when they result in damage to property or human health (Calantoni et al., 2022).

Although more could have been done earlier, the acceleration in prioritising environmental crimes in many parts of the world has been significant. In 2013 and even in 2017, environmental crime was considered only an emerging threat at the EU strategic level.

However, by 2021, this prioritisation became evident when the Serious Organised Crime Threat Assessment (SOCTA) included waste, pollution, and contamination crimes among the EU priorities (Alfaro Moreno, 2023). This shift, alongside the revised EU Environmental Crime Directive,[4] which mandates member states to criminalise 20 specific environmental offences, reflects the EU's attempt to form a unified front in tackling environmental crimes more efficiently. This acceleration is due to several factors, with the worsening global warming situation being one of them.

While green and critical criminology have consistently broadened the perception of wrongdoers or offenders, emphasising that environmental crimes are primarily committed by nation-states and transnational corporations in the pursuit of 'normal' business outcomes and which involve 'normal' business practices' (White, 2015), the discourse of many international security organisations remains predominantly focused on organised criminal groups. The inability or unwillingness to acknowledge the crimes of the powerful continues to cloud judgement of global community about climate change criminality.

The police narrative surrounding environmental crimes remains largely shaped by data and evidence emphasising the involvement of organised groups. While this is still accurate, in the long term, the police will have to confront the reality that the

---

[4] *Directive (EU) 2024/1203 of the European Parliament and of the Council of 11 April 2024 on the protection of the environment through criminal law.*

majority of environmental harms are perpetrated within legitimate sectors, even under the conditions of permits. This presents a significant challenge for the police. As noted by Amir Niknam, a police officer and innovation adviser from the Netherlands, it is not that the police are unaware of these issues, rather the absence of reporting leads to the exclusion of such offences from police statistics and discourse, resulting in a skewed understanding of environmental crimes. He further emphasised the necessity of quantifying the harm caused by corporations in order to disrupt the current pattern of neglecting crimes committed by powerful actors. It is also important to highlight that the Spanish SEPRONA, a branch of the military police exclusively dedicated to investigating environmental crimes, has a specialised unit composed of both police officers and scientists, including a biologist, physicist, veterinarian, and materials engineer. This unit is specifically tasked with quantifying the financial/economic impact of each environmental crime under investigation and producing clear, accessible reports supported by bibliographic references for use in criminal proceedings.

On the other hand, there is a prevalent assumption that responses to environmental crimes must be linear and progressive. However, there exists a paradox wherein poorly conceived solutions may, in fact, exacerbate the very problem they aim to address. Goulka and colleagues highlight this in the context of drug legislation and enforcement. The legalisation of cannabis production and consumption has long been advocated as a strategy to humanise the justice system and combat organised groups. Yet, if the legalising country or state has inappropriate climate and/or ecosystem for cannabis cultivation, it can lead to excessive water waste and increased carbon emissions due to the energy demands of indoor cultivation (Goulka et al., 2023). Similarly, technological solutions, intended to support sustainable transitions can inadvertently contribute to environmental degradation, such as the water shortages caused by the maintenance of computer systems. This aligns with one expert's observation, referenced frequently throughout the book, that responding to climate change, whether by the police or other institutions, inevitably involves trade-offs.

The legacy of environmental crime studies also shows the expanding range of actors, implicated in inflicting environmental harms. Sharif and Uddin (2021) in their description of the situation of factors contributing to environmental crimes in Bangladesh, describe them as exhibiting a spectrum from 'soft' to 'hard' characteristics. The former, such as citizens' harmful behavioural practices towards the environment, are quite common. In contrast, the latter include vehicular and industrial emissions, untreated waste dumping, river grabbing, encroachment, waste disposal, sand filling, illegal hill cutting, unauthorized brick kiln installations, illegal forest wood exploitation, wildlife poaching, cruelty against wildlife, wildlife smuggling, and pollution from ship-breaking yards. The authors observe that the economic aspect of environmental crime has two sides: one is profit maximization and the other is livelihood source (ibid.). The interplay between 'soft' and 'hard' characteristics of environmental crimes was very well echoed in the interview with a military/security expert from India who discussed the example of water shortages in the country:

> There is a lot of illegal drilling happening everywhere. There's also a lot of water tanker mafias which are operating illegally in many cities that are affected by water shortages. I'm not saying this happens all the time, but when there is a water shortage, like recently in Bangalore, which is one of the major metropolitan cities in South India. We have also seen in the past, like in Chennai, another city in the south, where that year Chennai faced close to day zero kind of situation, the taps ran out and the police had to step in. Actually, there were a lot of conflicts between different users of water. A lot of water was illegally taken from the suburban areas and the villages nearby and brought to the city. The villagers' consent was yet again not taken into account before the water was taken from there. There were a lot of criminal activities, including the ones committed by companies. This is not just happening in cities, but also farmers overuse and exploit the groundwater. (Interviewee 5)

The aforementioned section illustrates that the legacy of environmental crimes has broadened our understanding of offending, victimhood, and harm, including those harms that contribute and result from climate change. While this is valuable for advancing scholarly foundation on the impact of climate change on criminology and policing studies, practical examples on the ground show a far more complex and nuanced depiction of this reality.

The Meaning of the Ecocide Movement

Ecocide refers to significant destruction of the natural environment due to human activity. The Independent Expert Panel for the Legal Definition of Ecocide defines it as: 'unlawful or wanton acts committed with knowledge that there is a substantial likelihood of severe and either widespread or long-term damage to the environment being caused by those acts' (Independent Expert Panel for the Legal Definition of Ecocide, 2021). The ecocide movement does not focus solely on climate change but rather on ecosystem destruction, which is inherently linked to climate regulation, as pointed out by Jojo Mehta, ecocide expert from the United Kingdom, who also said that even if all emissions were halted immediately, the continued destruction of carbon sinks and climate-regulating ecosystems, such as mangroves, seas, and forests, would still contribute to the climate crisis.

Although still in its infancy, the emerging scholarship on climate criminality and climate offenders is increasingly being discussed alongside ecocide legislation and the coexisting movement. Ecocide is an example of harm with a long history of criminalisation attempts and has been at the forefront of various discussions since the Vietnam War. While it may still be considered a radical piece of legislation, it may not be viewed as such in the future, when the preservation of the planet is more widely accepted (Walters, 2023).

Like genocide, ecocide can be direct and indirect; Jojo Mehta, compared this legal development to a 'criminal version of health and safety for the planet'. In 2010 a proposal was submitted to the United Nations Law Commission by UK-based lawyer, Polly Higgins, to recognise ecocide as the fifth crime against humanity, included in the Rome Statute, and defined as: 'Ecocide is the extensive damage to, destruction of or loss of ecosystem(s) of a given territory, whether by human agency or by other causes, to such an extent that peaceful enjoyment by the inhabitants of that territory has been severely diminished' (Higgins et al., 2013).

Currently, the ecocide movement represents a broader shift in criminology and crime investigation as it challenges the necessity of 'intent' as a necessary crime component. Most ecocide is committed by corporations and often it is considered as an accident or collateral damage, which would give the defence the right to argue that corporations did not know what was happening or what consequences their action could bring about (Higgins et al., 2013). The ecocide movement and legislation bring the legal and criminological innovation to view crime also as a harm that results as an outcome of omission rather than intentional act.

The proposal to recognize ecocide as the fifth crime against humanity marks a critical step in defining harm and highlights the widespread environmental destruction that impacts humans, ecosystems, and nonhuman species alike. The establishment of ecocide as a crime is based on the principle of Earth stewardship, where the planet is considered to be 'held in trust', and humans are responsible for its proper care and management (White, 2018a, 2018b). Despite the noticeable achievements of the ecocide movement, a couple of experts, among them one of the criminologists from the United Kingdom, did not share the view that these will significantly help to curtail climate change criminality:

> About five years ago, when the ecocide proposal was announced, the journalists were really shocked when Philipson said that this won't do anything to prevent climate change. This is nothing to do with climate change. He was merely openly acknowledging what I've just said that the law is framed to deal with illegal or wanton acts which lead to widespread environmental damage, right? The purpose of the ecocide law is to deal with very specific acts that lead to long term damage. That's not the use of fossil fuels per se, because the use of fossil fuels is encouraged by law. I think the first thing that that we need to recognize is that probably the environmental problem we've got now is something that can't be dealt with just by dealing with a number of self-enclosed, self-defined criminal instances. (Interviewee 15)

Despite the criticism that the International Criminal Court (ICC) might not be the best institution for such responsibility, due to its predominantly Western-driven ideas and practices, as well as the lengthy processes involved, the ecocide movement's aim at the international level was intended to bring coherence and law harmonisation globally:

> We started off at the international level because we were aiming to create a kind of coherence. If you go piecemeal, one country at a time, you could end up with lots and lots of different things. Also, it's taking an extremely long time. Whereas actually by aiming for the International Criminal Court, what's actually happened is we've galvanised developments at the national level and they're all kind of having a certain level of coherence. (Jojo Mehta)

Ecocide and other environmental harms are primarily depicted as being committed by states, corporations, and organised criminal groups. However, Agnew argues that every day, seemingly trivial 'ordinary harms', such as living in a large home, driving to work, and consuming meat, also contribute to ecocide by increasing air, water, and soil pollution, destroying natural habitats and animal life, and depleting natural resources. Agnew (2012) further contends that societies with large economies and a long tradition of fostering materialistic attitudes are more likely to contribute to these daily, ordinary acts of ecocide. Thus, the ecocide movement can also be understood at the micro, individual level. The power of consumerist ideology and practices manifests in how certain forms of production and consumption become ingrained in everyday life, shaping our daily experiences and habits (White, 2015).

Setting aside the question of whether ecocide legislation achieves its intended outcomes, it nonetheless conveys three key messages. First, it underscores the necessity of harmonized and unified approaches to addressing environmental challenges, alongside sustained efforts to build momentum and raise awareness. However, regardless of its global orientation, the ecocide discourse ultimately converges on the individual level, inevitably raising uncomfortable questions about personal responsibility, consumer behaviour, and lifestyle choices.

Climate Crimes, Climate Offenders, Climate Litigation

There is a slowly emerging sub-category of very specific and closely tied with climate change crimes (Colantoni et al., 2022) such as trafficking of fluorinated greenhouses gases, misuse of sustainability funds, carbon credit and emissions trading scams, green energy investment fraud, fraudulent green lobby, illegal mining for renewable energy materials, greenwashing and fraudulent sustainability claims, or theft and vandalism of sustainability infrastructure. In the recent years the scholarly criminological literature is enriched with works that draw on the legacy of environmental crimes, the momentum around ecocide legislation, and introduces the language of 'climate crimes' or 'climate change criminality', 'climate offenders' meaning actions, or failures to act, by states and/or corporate organisations that contribute to global warming, deny the reality of climate change or its human causes, neglect to reduce greenhouse gas emissions, or respond to climate disruptions through unjust or overly securitised means (White, 2018a, 2018b; Kramer, 2020; Walters & South, 2020; Walters, 2023).

A good case that illustrates the emerging narrative of climate crimes is the commodification of drinking water being increasingly reported around the world and it was also discussed in my interview with Gorazd Meško, criminologist from Slovenia. Many communities around the world are increasingly facing acute water shortages intensified by the effects of heatwaves, creating a paradoxical situation. Multinational corporations exacerbated this crisis by extracting local water resources, processing them for commercial purposes, and selling them back to the affected populations. In Spain, this phenomenon is gradually invoking widespread criticism, with local communities mobilising to raise awareness and accusing these companies of exploiting natural resources for profit and effectively commodifying an essential public good.[5]

---

[5] https://www.theguardian.com/lifeandstyle/2024/nov/23/spanish-villages-people-forced-to-buy-back-owndrinking-water-drought-flood Accessed 25.11.2024.

The liability of legal entities plays a crucial role in addressing environmental crimes, as corporations and other legal entities are often the primary offenders related to air pollution and waste management regulations, which are the most common offences for which the liability of legal entities is explicitly addressed in environmental legislation (UNODC, 2024). This criminological narrative places a strong emphasis on corporations and states as the primary offenders responsible for climate change.

One of the most revolutionary aspects of the aforementioned EU Environmental Crime Directive was the redefinition of unlawfulness, which now recognises harmful actions even if they are conducted under the scope of a licence or permission. This is something that was extensively discussed with an environmental law expert from the Netherlands (Interviewee 9). This change specifically targets large corporations, aiming to dismantle their neutralisation techniques, as also discussed by Hallenberg (2021), which have long been used to downplay the environmental damage caused as part of profit-making processes.

> The article [Article 3] says that there is unlawfulness also when you comply with the conditions of a permit, on the condition that there is a serious breach of essential legal obligations. According to me this is the most important article of the whole directive. Why? In member States, we now have huge problems with serious pollution. Literally people dying today in Belgium, in the Netherlands. The whole problem is that when we watch interviews with the directors of some of these polluters on television, their answer is always the same - we comply with the conditions of the permit. (Interviewee 9)

In the same interview, it was also suggested that a good example of an emerging criminal practice that may increasingly require more police involvement is, for example, the case of emissions trading systems, which scholars have labelled as 'corporations' licence to pollute the planet' (Walters & Martin, 2013).

> We have seen a much stronger tendency to use market-based instruments with the evolution of environmental law. Emission trading is a market-based instrument. Reporting duties are market based. We get a sustainability due diligence directive.
> Market based instruments, due diligence. Much more flexible instruments are introduced. Yet very often these flexible instruments are also enforced through criminal sanctions. There may be a role for the police as well. (Interviewee 9)

Despite significant advances in environmental law, such as the previously discussed EU Environmental Crime Directive, the environmental legal expert (Interviewee 9) reflected on a sense of frustration among fellow environmental lawyers. Their frequent feeling of defeatism and disbelief regarding the ongoing environmental degradation led them to seek additional legal solutions, such as the development of ecological law. They believe that broader, more systemic ecological thinking might better address the scale of global environmental crises. By 2018, there were more than 1500 climate laws and policies implemented globally. Every country that signed the Paris Agreement had at least one law addressing climate change or the transition to a low-carbon economy, while 139 had comprehensive framework laws targeting climate mitigation and/or adaptation (Nachmany & Setzer, 2018). The mere introduction of additional laws may not be then the right solution to the problem. Another source of defeatism was expressed by the British critical criminologist who was concerned about the ability to curtail the scale and impact of corporate and white-collar crimes, which should be seen as the symptoms of an ever-expanding capitalist economy.

> I think most people who study corporate crime or white-collar crime want a different kind of society where it's not so easy for corporate crime to occur. What's changed in the outside world is a proliferation, certainly a proliferation of different forms of fraud and corruption, and different forms of environmental violence, violence against workers and so on. I mean, that's actually because the economy has expanded, and I think in the most recent period of global capitalism, the capacity for accumulation has expanded. The capacity for economies to accelerate creates new opportunities for fraud, corruption. (Interviewee 15)

Climate change will inevitably lead to policies and actions aimed at stopping greenwashing practices. As one of the expert interviewees pointed out, environmental crimes and harms will challenge the police regarding the prioritisation of environmental cases. The decision to investigate so-called carbon criminals, as well as harms and crimes that mostly result from the reciprocal relationship between corporate entities and governments (Kramer, 2020), will be difficult for the police to make, as indicated by David Lydon:

> I think in tandem with the ethical dimension or whether the state and the police, by virtue of them being the arm of the state, focus only on individuals in society rather than what the real problem is. Which you might argue, of course, are corporations and state actors who are the ones who some people might count them as carbon criminals. The people that are creating a lot of the problems and contributing to global warming, and all the shenanigans that go on with, carbon credits and all of that sort of stuff, you know, they're the problem. And yet, are the police then going to be more interested in dealing with individuals like you and I rather than big corporations? (David Lydon)

However, the mere presence of the police is not enough; the key issue is how enforcement is carried out. As highlighted in my interview with the British critical criminologist, a recurring conclusion, originally made by Sutherland, from studies on corporate crimes is that corporate entities often have the ability to bypass and circumvent mechanisms designed to ensure environmental responsibility. What is needed, therefore, are efficient interventions at the production cycle. According to the expert: 'We can't just set the bar with the law; it also needs to be enforced, and we need effective mechanisms to do so. I don't think we can criminalize our way out of this situation. We need a systemic change in production. It's a much deeper and broader issue—far beyond the scope of the police'. The expert supported this view be referring, for example, to the case of ban on single-use plastic bags, introduced in August 2017 in Kenya prohibiting their manufacture, sale, and use (Omondi & Asari, 2021).The discourse around climate crimes and climate change criminals will likely lead to investigations into state-facilitated climate change criminality, similar to Reece Walters' meticulous analysis of Australia's Royal Commission into National Disaster Arrangements following the country's summer bushfires of 2019/2020. Walter's research provides a detailed overview of how political inaction and omissions, combined with the prioritisation of fossil fuel economies, contributed to the devastating conditions for the bushfire ecocide. According to the author, this was a preventable environmental and human catastrophe (Walters, 2023).

The impact of climate change is inviting us to broaden our perception of harm further, and in so doing, is also creating new forms of legislation, it has particularly opened the gate to something that is now widely referred as climate litigation. For example, the 2021 ruling of the German Constitutional Court represents a historic milestone for what climate protection might be considered even beyond Germany. The ruling signals that the protection of climate is a constitutional obligation and the longer we wait and resists the greenhouse gas emissions the more likely it is that criminalisation process will include massive restrictions on freedom (Hong, 2023). In the Hong's summary of the sentencing decision, we read the following excerpt from the court's decision:

> We are driving towards a cliff. The longer we continue to drive at the current speed instead of slowing down towards climate neutrality, the harder we will have to hit the brakes later. Unless more far-sighted planning and action is taken now, a Corona-like societal "full braking" will be necessary in the near future.

A comprehensive literature review on climate litigation by Setzer and Vanhala (2019) revealed several interesting observations, also of criminological relevance. Although most of the literature relates to 76% of the total dataset from Global North jurisdictions, two valuable developments stand out in the review: (1) the growing diversity in disciplinary and interdisciplinary perspectives, and (2) the expanding range of actors involved in lawsuit (e.g. climate change activists, judges, fossil fuel divestment movement). The first wave of climate litigation tended to focus on high-profile cases and was dominated by legal scholars. While such cases undoubtedly helped establish precedents and highlight the global challenge of climate change, the review highlights the importance of using local and first-instance courts more effectively to address routine cases, such as allocation of carbon emissions.

It is argued in the review that these local and everyday disputes can also contribute to coherent national climate policy. More recent cases have included the impact of the litigation itself and its influence on climate change governance and regulation. This trend is coupled with the so-called human rights turn, meaning the inclusion of rights claim in climate change lawsuit. In addition, the growing role of science in the courtroom raises the question of how to effectively communicate scientific findings in judicial processes. However, the review shows that there is limited evidence regarding the true extent of the impact that climate litigation has, whether it drives tangible actions or merely raises awareness. Although, it seems that climate litigation is matter for lawyers and legal scholars, the accelerating effects of climate change and growing practice of climate litigation might call for a more prominent role for law enforcement in this process.

As climate-related crimes, each inherently linked to environmental harm, gradually emerge, there is a pressing need to develop the capacity to identify, investigate, quantify, and respond to them effectively. This process does not begin in a vacuum; the extensive body of scholarly literature, along with rapidly evolving investigative practices within policing and related sectors, offers valuable tools for anticipating and forecasting certain patterns of climate-related criminality. This is also an area where the growing contributions of green criminologists are becoming increasingly visible—somewhat distinct from the discussions presented in the following chapter.

References

Agnew, R. (2012). The ordinary acts that contribute to ecocide: A criminological analysis. In A. Brisman & N. South (Eds.), *Routledge international handbook of green criminology* (pp. 52–67). Routledge.

Alfaro Moreno, J. A. (2023). Towards an understanding of consequences of the prioritisation of environmental crime in the European Union. *Cuadernos de la Guardia Civil, 69*, 31–48.

Beck, U. (1992). *Risk society, towards a new modernity*. SAGE.

Berg, J., & Shearing, C. (2018). Governing-through-harm and public goods policing. *The Annals of the American Academy of Political and Social Sciences, 679*(1), 72–85.

Blaustein, J., Shearing, C., & Miccelli, M. (2024). Adaptive policing for a climate crisis. *Policing and Society: An International Journal of Research and Policy*, 1–13. https://doi.org/10.108 0/10439463.2024.2362713

Colantoni, L., Sarno, G. S., & Bianchi, M. (2022). Fighting environmental crime in Europe: An assessment of trends, players, and actions. Ambitus Project Report. Istituto Affari Internazionali (IAI). Retrieved from https://en.ambituseuropa.com Accessed 10.08.2024.

Farrall, S., Ahmed, T., & French, D. (2012). *Criminological and legal consequences of climate change*. Onati Institute for the Sociology of Law, Hart Publishing.

Goulka, J., Kang, S., Martin, A., Ensign, K., & Beletsky, L. (2023). When crises collide: Mapping the intersections of climate, pollution, crime and punishment. Northeastern University School of Law Research Paper No. 437, Northeastern University Law Review,15(2). Retrieved from https://ssrn.com/abstract=4453693 Accessed 15.02.2025.

Hallenberg, K. (2021). Crime, harm, and climate change nexus. In W. Leal Filho, A. M. Azul, L. Brandli, A. Lange Salvia, P. G. Özuyar, & T. Wall (Eds.), *Peace, justice and strong institutions* (pp.1–12). Springer.

Harrington, C., & Shearing, C. (2017). *Security in the Anthropocene. Reflections on Safety and Care*. Bielefeld: Transcript.

Higgins, P., Short, D., & South, N. (2013). Protecting the planet: A proposal for a law of ecocide. *Crime, Law & Social Change, 59*(3), 251–266.

Hong, M. (2023). Intertemporal freedom in the historic climate protection ruling of the German federal constitutional court. Verfassungsblog. Retrieved from https://verfassungsblog.de/ intertemporal-freedom-in-the-historic-climate-protection-rulingof-the-german-federal-constitutional-court/ Accessed 25.03.2024

Independent Expert Panel for the Legal Definition of Ecocide. (2021). Legal definition of ecocide. Stop Ecocide International. Retrieved from https://www.stopecocide.earth/legaldefinition Accessed 16.02.2025.

Kramer, R. C. (2020). *Carbon criminals, climate crimes*. Rutgers University Press.

Mythen, G. (2014). *Understanding the risk society. Crime, security and justice*. Red Globe Press.

Nachmany, M., & Setzer, J. (2018). *Policy brief global trends in climate change legislation and litigation: 2018 snapshot*. Grantham Research Institute on Climate Change and the Environment. Retrieved from https://www.lse.ac.uk/granthaminstitute/wpcontent/uploads/2018/04/Global-trends-in-climate-change-legislation-and-litigation-2018snapshot-3.pdf Accessed 15.10.2024.

Nellemann, C. (Editor in Chief); Henriksen, R., Kreilhuber, A., Stewart, D., Kotsovou, M., Raxter, P., Mrema, E., & Barrat, S. (2016). *The rise of environmental crime—A growing threat to natural resources peace, development and security. A UNEP-INTERPOL rapid response assess-*

ment. United Nations Environment Programme and RHIPTO Rapid Response–Norwegian Center for Global Analyses. Retrieved from www.rhipto.org Accessed 10.08.2024.

Omondi, I., & Asari, M. (2021). A study on consumer consciousness and behaviour to the plastic bag ban in Kenya. *Journal of Material Cycles and Waste Management, 23*(2), 425–435.

Setzer, J., & Vanhala, L. C. (2019). Climate change litigation: A review of research on courts and litigants in climate governance. *Wiley Interdisciplinary Reviews: Climate Change, 10*(3), e580.

Sharif, S. M., & Uddin, M. K. (2021). Environmental crimes and green criminology in Bangladesh. *Criminology & Criminal Justice, 23*(3), 490–510.

United Nations Office on Drugs and Crime (UNODC). (2024). *Global analysis on crimes that affect the environment—Part 1: The landscape of criminalization.* United Nations Publication.

Walters, R. (2023). Ecocide, climate criminals and the politics of bushfires. *The British Journal of Criminology, 63*(2), 283–303.

Walters, R., & Martin, D. (2013). Crime and the commodification of carbon. In R. Walters, D. Solomon Westerhuis, & T. Wyatt (Eds.), *Emerging issues in green criminology: Exploring power, justice and harm* (pp. 93–107). Palgrave Macmillan.

Walters, R., & South, N. (2020). Power, environmental harm and the threat of Global Ecocide. In L. Copson, E. Dimou, & S. Tombs (Eds.), *Crime, harm and the state* (pp. 141–178). The Open University Press.

White, R. (2015). Climate change, ecocide, and crimes of the powerful. In G. Barak (Ed.), *The Routledge international handbook of the crimes of the powerful* (pp. 211–222). Routledge.

White, R. (2018a). *Climate change criminology.* Bristol University Press.

White, R. (2018b). Ecocide and the carbon crimes of the powerful. *The University of Tasmania Law Review, 37*(2), 96–117.

Chapter 4
Feeling the Heat

High Temperature and Crime Nexus

The most significant shifts in extreme weather due to climate change involve changes in the frequency and intensity of temperature extremes. While cold extremes are becoming less frequent and less severe, heat extremes are on the rise (Clarke et al., 2022). The rise in temperatures is the most significant manifestation of climate change, with other impacts such as water shortages, droughts, shifting precipitation patterns, and changing wind patterns being secondary consequences, that is, flooding and storms. An excellent depiction of what it means to be affected by high temperatures can be found in the climate fiction novel *Ministry for the Future* by Kim Stanley Robinson, published in 2020, which vividly describes human suffering caused by prolonged exposure to extreme heat. Another book, *The Heat and the Fury*, published in September 2024 and written by environmental journalist Peter Schwartzstein, explores this topic from a journalistic and documentary perspective. Unsurprisingly, one of the most frequently raised question about climate change is whether higher temperatures will affect human behaviour, and if so, how? This was excessively discussed with Interviewee 10, a heat-crime scholar from the United States.

> Heat is one of the most obvious consequences of climate change. I also think it's true that of all of the effects of all of the things that climate change will shift in our weather systems, heat is the thing that is likely to adversely affect the largest number of people around the world. (Interviewee 10)

The impact of increased temperatures, for example, on mortality is already widely established in the epidemiological literature (see Ballester et al., 2023), and this trend is gradually evident in establishing heat-crime nexus (Hipp et al., 2004; Ranson, 2014; Burke et al., 2015; Xu et al., 2020; Xu et al., 2021; Kubo et al., 2021). This scholarship about heatwaves forms a bigger picture on how weather conditions can influence people's behaviour. There is also evidence that pollution has an impact on violent crime or unethical behaviour (Burkhardt et al., 2019; Gong

© The Author(s) 2025
A. Matczak, *Adapting to Climate Change in Modern Policing*, SpringerBriefs in Criminology, https://doi.org/10.1007/978-3-031-97510-3_4

et al., 2020; Herrnstadt et al., 2021). Although there is still a long way to go in establishing exact causation, the simultaneous development of 'attribution science'—the study of the relationship between climate change and weather events and impacts—helps strengthen this area of research. What is known at this stage is that heat diminishes self-control, adversely affects mood, escalates aggression, and puts additional strain on cognitive abilities (Behrer & Bolotnyy, 2024). These psychological effects both contribute to and coexist with broader societal tensions, offering a valuable lesson for criminology to integrate psychological insights alongside sociological theories of crime for a more balanced understanding of the impact of climate change on the landscape of crimes and harms. Interviewee 10 elaborated on this as follows:

> I think this is driven by the sort of physiological response that people have to heat. It makes you grumpy. It reduces your cognitive control. When you're put in a situation that on a non-hot day might not turn into violence, the increase in aggression and reduction in control result in violence. I think that literature is quite robust. There are mediating factors. There's an interesting paper that was published a couple of months ago from Mexico that shows that access to alcohol plays a really important role in mediating that relationship as well. People typically drink a lot more on hot days, which exacerbates the impact of heat. (Interviewee 10)

The scholarly findings of the impact of heatwaves on human behaviour have already been observed in police operations, a topic that was also extensively discussed by Alexandra Jones who is based in the Netherlands:

> Many people live in the Hague in flats. They have no balcony, they have no air conditioning, or very little, no adequate air conditioning, and typically inhabited by large families. If you live on the ground, on a top floor with a flat roof, and you have 4 or 5 children in your house and three bedrooms, things get very awkward. You see this in the summer holidays when the temperature rises to get young people going out into the street because it's simply too hot to stay inside and they start vandalising things. If it becomes a stress rather than a shock you can deal with the heatwave that lasts for days because there's an end in sight. Oh, God. What if it lasts for months? (Alexandra Jones)

Societies experience climatic variables in real time and respond to both short-lived and long-term events. For example, if high temperatures increase the likelihood of riots in a city, even if extreme heat is experienced only for a few hours, then this is important for understanding climate impacts because the frequency of these momentary events may change if the distribution of daily temperatures changes (Burke et al., 2015).

There are already examples of police involvement in actions related to heat or heatwaves. Italian Police in Portofino was asked to monitor the illegal setting up of air conditioning in the city as the new regulation banned the installation of the cooling units on the traditionally known pastel-coloured Portofino historic buildings.[1] Greek Police are involved in searching for missing tourists who miscalculated the impact of summer heat on their hiking plans.[2] These are still relatively low-threshold policing activities but the bigger question to ask is whether higher temperatures increase people's propensity for conflict, and then in consequence to commit crimes.

---

[1] https://www.theguardian.com/world/article/2024/aug/12/portofino-air-con-crackdown-neighbours-italy Accessed 13.08. 2024.

[2] https://www.skynews.com.au/world-news/police-ramp-up-search-for-missing-dr-michael-mosley-whodisappeared-on-greek-island-during-extreme-heatwave/news-story/611ad0d5fa149215a999d9e010fe4499 Accessed 13.08.2024.

There is also some evidence suggesting an increased demand for police services during the warm season (Williams et al., 2020), and that heatwaves might bring about psychological and cognitive consequences for police officers, as well as judges, and influence their decision-making processes about arrests and sentencing (Behrer & Bolotnyy, 2024). This line of research can assist police services in developing climate change adaptation strategies, managing budgets, staff rotation, ensuring resilience against the rising frequency, severity, and duration of extreme heat periods. It is likely only a matter of time before police agencies begin integrating weather data into their operational planning.

In the coming decades, extreme weather events linked to global warming will challenge the sustainability, resilience, and adaptability of various systems and sectors, which are not traditionally a typical domain of criminological research. For example, another dimension in the discussion of the heat-crime nexus is the impact of heatwaves on the informal economy, construction activities, safety deficiencies, and negligence in safety practices, something that was also discussed with the security expert from India:

> We have a huge informal economy [in India] like the construction sector and others. Heatwave is something that affects the labour productivity. The police generally have to come in to ensure that labour rights are protected. These are all expectations. I don't mean to say that it actually happens on the ground. These are all things that the police are also very much involved in. (Interviewee 5)

The current research findings confirming the relationship between higher temperatures and people's proclivity to conflict might be mitigated by the simultaneous emergence of climate change-adjusted infrastructure. There are already debates about geographical areas, cities that are more climate change-proof.[3] As observed by Clifford Shearing, it is also possible to imagine that with the accelerating effects of higher temperatures on the planet, communities will adapt their households and public spaces.

> We may be moving to that kind of existence where we are living in enclosed spaces and with domes over everything. I suspect we'll see more of that kind of infrastructure emerging. So, we'll create our own climate. So that's what these underground areas in Canada are about. You're creating your own climate. So, this move towards spaces that have more compatible climates for humans. You know maybe a direction we'll see emerging because, a lot of farming is already going under cover. So, if you look around rural or urban area, I suspect you will find increasing evidence for this. I suspect this is the case in the Netherlands though I haven't been there for a long time. (Clifford Shearing)

Extreme heat can also negatively affect road users by increasing fatigue, inattentiveness, or reckless crossing behaviour. A study in Boston found that weather conditions involving heat or rain were associated with road injuries and property damage. A potential intervention could involve testing whether the behaviour of drivers, cyclists, and pedestrians during adverse weather can be modified through devices such as vehicle dashboards, mobile phone alerts, and portable road signs (Nazif-Munoz et al., 2021).

---

[3] https://www.nationalgeographic.com/premium/article/climate-change-havencities?cmpid=org=n gp::mc=social::src=instagram::cmp=editorial::add=ig20240730environmentclimatehavenscitiesu spremiumhedcard&linkId=529004162 Accessed 4.08.2024.

Despite the growing body of research on the impact of high temperatures on human behaviour and wellbeing, what makes this particular consequence of climate change especially dangerous is its regularity. Unlike extreme weather events, exposure to high temperatures has become a routine, everyday experience. In traditionally cooler regions, higher temperatures may even be greeted with gratitude and joy. The deceptive and insidious nature of rising temperatures allows them to quietly infiltrate daily life, concealing their influence on human behaviour and wellbeing, as pointed out by the heat-crime expert from the United States, who said that: 'there's an important distinction between disasters or really severe shocks and the run of the mill day to day exposure to high temperature'—meaning that one of the key risks associated with sustained high temperatures is that it does not command immediate attention or urgency. Unlike disasters, which are highly visible and elicit strong adaptive responses, prolonged heat exposure tends to be overlooked. This is not necessarily because people consciously ignore it, but because their attention is occupied by other pressing concerns. The interviewee then gave an example of workplace injuries. Many incidents that are likely caused by extreme heat occur in already hazardous environments. Workers in such settings may not recognise heat as a primary risk factor because their focus is on other dangers inherent to their jobs. Roofers, for example, are especially vulnerable, as they work at high elevations under direct sunlight. They are more likely to fall on hot days, yet they may not attribute this risk to heat itself. Instead, their primary safety concerns revolve around structural hazards, such as unstable ladders or falling materials, rather than temperature-related risks. This contrast highlights how differently people respond to everyday heat exposure compared to disasters. According to the expert:

> Which makes the effects of heat particularly insidious, because you're not doing things to protect yourself from the heat, necessarily. You're doing things to protect yourself from falling or other things. As opposed to, say, a disaster where that's a highly salient event, it attracts attention and consequently attracts sort of adaptive responses or responses that are intended to minimise the negative consequences of the disaster. I think that people respond very differently because of the different salience of run of the mill heat versus disasters on the research side. (Interviewee 10)

In the discussion on the impact of higher temperatures on human behaviour, one important distinction that needs to be made is between interpersonal and intergroup conflicts. For example, the impact of climate change, and depleting space for farming, as one of the root causes for the herder-farmer conflict in Nigeria (Egbuta, 2018), in Kenya (see Parenti, 2011). Burke et al. (2015:5) define the former as 'conflicts between individuals, which include various acts commonly described as crime, such as assault, rape, and robbery, as well as other types of conflict that may not necessarily be criminal, such as violence in sporting events, road rage, and violent acts by police' and the latter as '*conflicts between collections of individuals, such as organised political violence, civil conflicts, wars, riots, and land invasions*'. The authors also mention two other types of potential conflicts: intrapersonal violence (e.g. suicides) and institutional breakdown and population collapse. This distinction was eloquently described by the heat-crime expert:

> Separate from that interpersonal literature, there's a whole distinct literature on intergroup conflict, and that includes both state and non-state actors. That literature tends to find that

heat increases inter-group conflict both between state actors and between non-state actors. There's robust evidence for this from Africa, from Southeast Asia, from the Middle East. The mechanisms there are much as you might expect, are much more complex than in the interpersonal literature and have to do with the role of heat in changing agricultural output and in changing the resources available both to individuals who want to engage in conflict, to sort of supply that conflict, as well as the incentives to engage in conflict, to control resources. (…) In general, I find that literature quite convincing, although I do think that the specific mechanisms vary from place to place, and I think that the interpersonal results are far more generalisable. People everywhere tend to have the same physiological response to heat that's mediated by access to air conditioning and infrastructure. There are things that can mitigate that relationship. But in general, the physiological response to heat is the same everywhere. Whereas the sort of group level response to heat exposure and whether that leads to conflict is highly dependent on the institutional context and so varies a lot across settings. (Interviewee 10)

In Peter Schwartzstein's book *The Heat and the Fury: On the Frontlines of Climate Violence* in the chapter on farmer-herder conflicts in West Africa, the author empha-sises that it is erratic nature of the rainfall, 'that uncertainty, that erratic deviation from the known, that is principally responsible for bringing farmers and pastoralists into uneasy contact', as something that was repeatedly said in his interviews across Senegal, Sudan, Burkina Faso, Kenya, and Mauritania (Schwartzstein, 2024:149).

On the other hand, there has always been a caution in suggesting a straightfor-ward connection between environmental degradation and environmentally induced conflicts (Homer-Dixon, 1991, 1994) due to the complexity and variety of conflicts. These conflicts can be subnational and low intensity (Bachler et al., 1996). In addi-tion, methodological and conceptual limitations of this type of research projects are meticulously delineated by Burke et al. (2015). The argument that environmental scarcity led to conflicts has been challenged by another line of empirical research showing how environmental degradation often provides an opportunity for group cooperation (Hauge & Ellingsen, 2001), demonstrating that it is resource abundance rather than scarcity that starts conflicts (Berdal & Malone, 2000). The research on the exposure to daily heatwaves, which produces more invisible implications, is dif-ferent from the disaster studies that, quite salient in nature, show people's propen-sity for action and cooperation, and suggests that these events can also improve social cohesion and collective efficacy (see Townsend et al., 2015).

Yet, there is substantial evidence confirming that higher temperatures negatively impact human behaviour and decision-making processes, contributing to increased interpersonal conflicts and harms. However, as Harrington and Shearing (2017:129) note, security in the Anthropocene is not only about predicting the 'hot spots' of climate conflicts but also about investigating solidarity and cooperation during cri-ses. While humanity may adapt to rising temperatures by developing climate-proof private and public spaces, such as the positive effects of green spaces on crime prevention, as seen in Goulka et al. (2023), and thus mitigate the risk of interper-sonal violence, addressing intergroup conflicts during adverse weather conditions presents a far more complex challenge. Furthermore, lessons from India highlight the importance of advancing research on how to better measure and predict heat-waves. As indicated pointedly in one of my interviews, relying on airport-based temperature data may not accurately reflect people's actual heat exposure in various environments, particularly in urban versus rural settings (Raval et al., 2018).

Climate Refugees and the Crimes of 'Otherness'

Another frequently raised question about the impact of climate change, manifested in adverse weather conditions and disasters, is the issue of growing land areas that are uninhabitable and how this pushes people to move within and beyond countries.

According to the data published by the Internal Displacement Monitoring Centre (IDMC), an NGO in Geneva, the overall global number of internally displaced people (IDPs) continues to rise at a significant pace. At the end of 2022 there were all-time high 71.1 million people living in internal displacement worldwide, a 20% increase in a year and the highest number ever recorded, and 47% increase since 2013 when disaster data were first available (IDMC, 2023). Almost three-quarters of the world's internally displaced people (IDPs) reside in just 10 countries: Syria, Afghanistan, the Democratic Republic of the Congo (DRC), Ukraine, Colombia, Ethiopia, Yemen, Nigeria, Somalia, and Sudan. In 2022, while conflict and violence[4] were the reason for 62.5 million people living in displacement across 65 countries and territories, disasters accounted for a further 8.7 million across 88 countries and territories, which is a 45% increase in the number of people internally displaced by disasters since 2021.

Climate change exacerbates these issues. The IDMC highlights the emerging intersectionality of adverse weather conditions, disaster, conflict, and displacement, which can then lead to long-distance migration and loss of social capital. By the end of 2023, nearly 3 in 4 forcibly displaced people were living in countries with high-to-extreme exposure to climate-related hazards (UNHCR, 2023). Although the concept of climate or environmental refugees is not legally recognised, there is growing discussion around their classification of their status, the interconnection with other experiences and hardships (such as armed conflict), and their specific needs and experiences, and various country-wide initiatives to accommodate them. However, unlike those fleeing armed conflicts, climate refugees, often referred as the 'forgotten victims of climate change', cannot currently claim asylum based on climate-related reasons alone (Bellizzi et al., 2023). They are now a valuable source of information about micro-level, individual vulnerability to climate change in the form of now well-documented loss of social cohesion and sense of identity (Di Giorgi et al., 2020).

It should be made explicitly clear that climate refugees themselves are not the problem; rather, the issue lies in the risks associated with their displacement. These risks are connected to their inherent vulnerability and include exploitation by criminal networks, the precarious conditions under which they live and work, and the societal and institutional responses to their movement and presence—responses that can often be harmful, abusive, or criminogenic. In essence, it is the combination of their

---

[4]Conflict and violence are operationalised in the report as: (1) international armed conflict, (2) noninternational armed conflict, (3) communal violence, (4) crime-related violence, (5) civilian-state violence, and (6) other forms of violence.

marginalised status and the societal perception of their 'otherness' that can contribute to crime and victimisation. This issue was also highlighted in several expert interviews.

> There's a lot of evidence to show that many of these farmers [in India] who are affected by repeated droughts are forced to move temporarily to cities for jobs, sometimes permanently. Sometimes they leave the families behind. Some of the states, the more developed states like Kerala, Karnataka, have huge numbers of migrant labourers now. Some special provisions are being made to ensure that they're not exploited. These are all recent things. For a long time, there was widespread exploitation, labour rights being violated. There is greater recognition of that as well. (Interviewee 5)
> I think we should be proactive towards politicians and make sure that they don't create a climate of hate, because if they do, we [the police] have to clean up the shit. (Amir Niknam)

The marginalised communities at a heightened risk of environmental/ecological disasters are now referred to as 'sacrifice zones'—fenceline communities of low-income residents living immediately adjacent to heavily intoxicated, polluted industries or military bases (see Lerner, 2012). A similar observation can be made regarding the expanding areas of the planet that are becoming uninhabitable due to climate change. These emerging 'sacrifice zones' may be as climatically inhospitable and socially unwelcoming in the place of origin as they are in the host or destination regions to which displaced populations may migrate.

Climate, Migration, and Conflict: The Radicalisation Nexus

At the extreme end of the evolving crime and harm landscape shaped by the accelerating effects of climate change is also the question of its impact on the emergence, formation, and activity of radical and terrorist groups. As with the previously discussed topics, there is a growing body of literature on this subject depicting climate change as an increasingly direct and indirect significant driver of terrorism (Feitelson & Tubi, 2017; Silke & Morrison, 2022; Mavrakou et al., 2022; Dmello & Neudecker, 2024). In their introduction to a special issue on Climate Change and Terrorism, Silke and Morrison (2022) reference David Rapoport's Four Waves of Modern Terrorism theory (2004) to discuss the addition to how modern terrorism has historically unfolded in four distinct waves. So far, each of these waves has emerged as an unintended consequence of broader forces closely tied to major global and geopolitical events and processes. They suggest that climate change may play a pivotal role in shaping a potential fifth wave of terrorism. In the aforementioned book *The Heat and the Fury* by Peter Schwartzstein, there is a compelling account in one of the first chapters of how water crises contributed to the rise of ISIS (Islamic State) in Iraq. In river-adjacent villages such as Huiesh, where farmers had relatively consistent access to the waters of the Tigris that allowed them to keep reasonably stable crop cultivation, very few inhabitants aligned themselves with jihadist groups. By contrast, the situation in more remote communities, located beyond the reach of the river and its canal systems, was markedly different (Schwartzstein, 2024).

While the scholarly literature on this issue unanimously asserts a positive relationship between climate change and terrorism, Mavrakou et al. (2022) argue that the field needs to advance by empirically supporting these claims and developing more elaborate and nuanced analytical tools to examine this relationship in greater depth. This observation as also echoed in one of the expert interviews:

> There is some anecdotal evidence to say that when the flood occurs, and if the government is not able to provide relief, these radical groups step in and fill the void. For instance, parts of Kashmir or Pakistan are always used as an example for the 2010 floods to show how radical extremist organisations use the floods as an opportunity to gain legitimacy and increase the recruitment drive. There's a scholar in Pakistan who did a study on the same issue, which is often used by climate and security scholars as a proof. She interviewed a lot of the locals who are affected by the flood. They go with these radical groups maybe for a month or two because they need money and then they leave. They don't really stay for a really long time. They may just take the help and leave. (Interviewee 5)

The question of the impact of climate change on radicalisation and terrorism is an extension of the broader question of how changing weather patterns influence human behaviour and movement. Although terrorism traditionally falls within the domain of political science, geopolitics, and/or security studies, it is also of key concern for certain dedicated police branches and the debate over the crime-terrorism nexus. Similarly to the topic of climate refugees, superficial discussions about climate change and terrorism create risk fostering 'compulsive climatic determinism'—a knee-jerk reaction to the perceived 'yet-to-come threat' (Telford, 2023). Telford warns against determining causal links between factors like rainfall variability, temperature shifts, food insecurity, displacement, and violent conflict—all of which have complex interconnections. Oversimplifying these relationships could potentially lead academic and practice communities to suggest simplistic solutions to the inherently complex and multifaceted nature of climate change and the multiple ways it manifests. That said, disregarding this line of discussion and failing to engage with the relevant literature risks oversimplifying and underestimating this specific consequence of climate change. Unlike the preceding chapter, where the contributions of criminological and policing scholarship are evident, the issues addressed in this chapter have seldom attracted any meaningful criminological attention, which I consider as a particularly significant omission.

References

Bächler, G., Böge, V., Klötzli, S., Libiszewski, S., & Spillmann, K. R. (1996). *Kriegsursache Umweltzerstörung: Ökologische Konflikte in der Dritten Welt und Wege ihrer friedlichen Bearbeitung [Environmental destruction as a cause of war: Ecological conflicts in the third world and ways of their peaceful resolution]*, Vol. 1. Rüegger.

Ballester, J., Quijal-Zamorano, M., Méndez Turrubiates, R.F. et al. (2023). Heat-related mortality in Europe during the summer of 2022. *Nature Medicine, 29*(7), 1857–1866.

Behrer, A. P., & Bolotnyy, V. (2024). Heat and law enforcement. *The Proceedings of the National Academy of Sciences Nexus, 3*(5), 1–8.

Bellizzi, S., Popescu, C., Panu Napodanu, C.M., Fiamma, M., & Cegolon, L. (2023). Global health, climate change, and migration: The need for recognition of "climate refugees". *Journal of Global Health, 13*, 13001. DOI: 10.7189/jogh.13.03011

Berdal, M., & Malone, D. M. (Eds.). (2000). *Greed and grievance: Economic agendas in civil wars*. Lynne Rienner Publishers.

Burke, M., Hsiang, S. M., & Miguel, E. (2015). Climate and conflict. *The Annual Review of Economics, 7*, 577–617.

Burkhardt, J., Bayham, J., Wilson, A., Carter, E., Berman, J. D., O'Dell, K., Ford, B., Fischer, E. V., & Pierce, J. R. (2019). The effect of pollution on crime: Evidence from data on particulate matter and ozone. *Journal of Environmental Economics and Management, 98*(C), 102267. https://doi.org/10.1016/j.jeem.2019.102267

Clarke, B., Otto, F., Stuart-Smith, R., & Harrington, L. (2022). Extreme weather impacts of climate change: An attribution perspective. *Environmental Research: Climate, 1.* 012001. DOI 10.1088/2752-5295/ac6e7d

Di Giorgi, E., Michielin, P. And Michielin, D. (2020). Perception of climate change, loss of social capital and mental health in two groups of migrants from African countries. Ann Ist Super Sanità, 56(2), 150–156.

Dmello, J. R., & Neudecker, C. H. (2024). Monsoon marauders and summer violence: Exploring the spatial relationship between climate change and terrorist activity in India. *Journal of Applied Security Research, 19*(3), 351-376.

Egbuta, U. (2018). Understanding the herder-farmer conflict in Nigeria. *Conflict Trends, 3,* 40–48. Retrieved from https://www.researchgate.net/profile/Darlington-Tshuma-2/publication/330441146_Looking_Beyond_2023_What_Next_After_Zimbabwe%27s_Parliamentary_Quota_System/links/5c403d1692851c22a37ae559/Looking-Beyond-2023-WhatNext-After-Zimbabwes-Parliamentary-Quota-System.pdf#page=41 Accessed 6.03.2024.

Feitelson, E., & Tubi, A. (2017). A main driver or an intermediate variable? Climate change, water, and security in the Middle East. *Global Environmental Change, 44*, 39–48.

Gong, S., Lu, J. G., Schaubroeck, J. M., Li, Q., Zhou, Q., & Qian, X. (2020). Polluted psyche: Is the effect of air pollution on unethical behaviour more physiological or psychological? *Psychological Science, 31*(8), 1040–1047.

Goulka, J., Kang, S., Martin, A., Ensign, K., & Beletsky, L. (2023). When crises collide: Mapping the intersections of climate, pollution, crime and punishment. Northeastern University School of Law Research Paper No. 437, *Northeastern University Law Review, 15*(2). Retrieved from https://ssrn.com/abstract=4453693 Accessed 15.02.2025.

Harrington, C., & Shearing, C. (2017). *Security in the Anthropocene. Reflections on Safety and Care.* Bielefeld: Transcript.

Hauge, W., & Ellingsen, T. (2001). Causal pathways to conflict. In P. F. Diehl & N. P. Gleditsch (Eds.), *Environmental conflict* (pp. 36–57). Westview Press.

Herrnstadt, E., Heyes, A., Muehlegger, E., & Saberian, S. (2021). Air pollution and criminal activity: Microgeographic evidence from Chicago. *American Economic Journal: Applied Economics, 13*(4), 70–100.

Hipp, J. R., Bauer, D. J., Curran, P. J., & Bollen, K. A. (2004). Crimes of opportunity or crimes of emotion? Testing two explanations of seasonal change in crime. *Social Forces, 82*, 1333–1372.

Homer-Dixon, T. F. (1991). On the threshold: Environmental changes as causes of acute conflict. *International Security, 16*(2), 76–116.

Homer-Dixon, T. F. (1994). Environmental scarcities and violent conflict: Evidence from cases. *International Security, 19*(1), 5–40.

IDMC. (2023). Internal displacement and food insecurity. Norwegian Refugee Council.Retrieved from https://api.internaldisplacement.org/sites/default/files/publications/documents/IDMC_GRID_2023_Global_Report_on_Internal_Displacement_LR.pdf Accessed 4.03.2024.

Kubo, R., Ueda, K., Seposo, X., Honda, A., & Takano, H. (2021). Association between ambient temperature and intentional injuries: A case-crossover analysis using ambulance transport records in Japan. *Science of the Total Environment, 774*, 145770.

Lerner, S. (2012). *Sacrifice zones: The front lines of toxic chemical exposure in the United States.* The MIT Press.

Mavrakou, S., Chace-Donahue, E., Oluanaigh, R., & Conroy, M. (2022). The climate change–Terrorism nexus: A critical literature review. *Terrorism and Political Violence, 34*(5), 894–913.

Nazif-Munoz, J. I., Martinez, P., Williams, A., & Spengler, J. (2021). The risks of warm nights and wet days in the context of climate change: Assessing road safety outcomes in Boston,

USA, and Santo Domingo, Dominican Republic. *Injury Epidemiology, 8*(47), 1–9. https://doi. org/10.1186/s40621-021-00342-w

Parenti, C. (2011). *Tropic of chaos. Climate change and the new geography of violence.* Nation Books.

Ranson, M. (2014). Crime, weather and climate change. *Journal of Environmental Economics and Management, 67*(3), 274–302.

Rapoport, D. C. (2004). The four waves of modern terrorism. In A. K. Cronin & J. M. Ludes (Eds.), *Attacking terrorism: Elements of a grand strategy* (pp. 46–73). Georgetown University Press.

Raval, A., Dutta, P., Tiwari, A., Ganguly, P. S., Sathish, L. M., Mavalankar, D., & Hess, J. (2018). Effects of occupational heat exposure on traffic police workers in Ahmedabad, Gujarat. *Indian Journal Of Occupational and Environmental Medicine, 22*(3), 144–151.

Robinson, K. S. (2020). *The ministry for the future.* Orbit.

Schwartzstein, P. (2024). *The heat and the fury: On the frontlines of climate violence.* Island Press.

Silke, A., & Morrison, J. (2022). Gathering storm: An introduction to the special issue on climate change and terrorism. *Terrorism and Political Violence, 34*(5), 883–893.

Telford, A. (2023). Where to draw the line? Climate change-conflict-migration-terrorism casual relations and a contested politics of implication. *Environmental Science and Policy, 141*, 138–145.

Townshend, I., Awosoga, O., Kulig, J., & Fan, H. (2015). Social cohesion and resilience across communities that have experienced a disaster. *Natural Hazards, 76*(2), 913–938.

UNHCR. (2023). *Global trends. Forced displacement in 2023.* Retrieved from https://www.unhcr. org/global-trends-report-2023 Accessed 28.10.2024.

Williams, A., McDonogh-Wong, L., & Spengler, J. D. (2020). The influence of extreme heat on police and fire department services in 23 U.S. cities. *GeoHealth, 4*, e2020GH000282. https:// doi.org/10.1029/2020GH000282

Xu, R., Xiong, X., Abramson, M. J., Li, S., & Guo, Y. (2020). Ambient temperature and intentional homicide: A multi-city case-crossover study in the US. *Environment International, 143*, 105992.

Xu, R., Xiong, X., Abramson, M. J., Li, S., & Guo, Y. (2021). Association between ambient temperature and sex offense: A case-crossover study in seven large US cities, 2007–2017. *Sustainable Cities and Society, 69*, 102828.

Chapter 5
Policing Civil Disobedience

Policing Environmental and Climate Demonstrations

While writing this book, it became clear that one of the most hotly debated topics surrounding the impact of climate change is the police response to environmental and climate demonstrations. Blaustein and colleagues also observe that this is the topic that gets most attention among policing scholars who study climate change (Blaustein et al., 2023). Unsurprisingly, this issue was also extensively covered in the expert interviews featured in the book, albeit with varying opinions. In the introduction, I referenced Alexandra Jones' quote on 'hurt distribution' to illustrate how the climatisation of police and policing is intertwined with the broader burden and costs of climate change, which will ultimately affect us all. Regardless of differing views on the role and impact of these demonstrations, it is evident that they play an important role in communicating and extending the 'hurt', and the police will increasingly find themselves involved in these dynamics. Amir Niknam emphasised the key role of such demonstrations as a necessary component for change in the bigger picture, stating:

> The main message is that some of our institutions are no longer working. If you look throughout history, when institutions were obsolete or they no longer had function, there were protests. It's easy to look away when the pain is not yours. The protest makes it impossible to look away because the pain becomes more apparent. It's here and now. It is a necessary component to change the institutions so that they become new and relevant, and we as a society can move forward again. In that sense, protests have a very clear function and a purpose. The fact that they are painful also has a purpose. If you look at the purpose it appears that it is very close to the purpose that we have as police, which is to help society. Even though we have different roles to play, we have different roles to play within the same, bigger picture. (Amir Niknam)

In like manner, according to Clifford Shearing, environmental and climate demonstrations are not an anomaly from a historical perspective; civil disobedience has always been a significant part of police's responsibilities:

© The Author(s) 2025

A. Matczak, *Adapting to Climate Change in Modern Policing*, SpringerBriefs in Criminology, https://doi.org/10.1007/978-3-031-97510-3_5

> I think there's nothing new about these protests. Protests about everything have been going on for a long time. Organising public events, managing public events and managing conflicts between groups – there's a lot of skill that police already have.

> Public order policing that is may well be reshaped. It's people fighting with each other, chaining themselves, stopping traffic, doing all this sort of stuff. It's in a sense a continuity thread. (Clifford Shearing)

In recent decades, there have been significant changes in the nature of protest, with an intensification of demonstrations over environmental issues that have posed new challenges to policing organisations. These challenges stem from the diversity of protesters, ranging from families with children to pensioners, and the variety of tactics employed, such as gluing oneself to the street or protesting at cultural events. The composition of these groups, including well-educated individuals, the elderly, and artists, makes them more relatable and challenges the traditional profile of 'police property' (Reiner, 2009). What is meant by the 'police property' in criminology and policing studies is individuals or groups that are traditionally viewed by law enforcement as suspects, frequent detainees, or those under constant police surveillance and control. It implies a stereotype of people who are expected to have repeated encounters with the police, often due to their socio-economic background, lifestyle, or perceived criminal propensity.

What is seldom acknowledged and missing in the scholarly literature is that there are also police officers, in their personal capacity, who also participate in these protests. I still vividly recall an interview with a police service employee, interviewed for a different project, who reflected on a striking moment when attending one of the climate demonstrations: one side of the protest were the active-duty police officers trained by my interviewee, and on the other, there were other colleagues from the police protesting alongside their families.

A comparative study of climate strikers in six different cities (Brighton, London, Montreal, New York, New Haven, Stavanger) conducted in September 2019 demonstrated protestors' different degrees of knowledge about climate change, different set of actions taken to address it in their daily lives, as well as a wide range of reasons for participating in the strikes (Martiskainen et al., 2020). The 'professionalisation' of environmental protests includes the emergence of full-time, well-trained protesters, who adhere to certain codes of conduct (Button et al., 2002). They not only resemble an everyday street-level resistance to environmental harms, they also write press releases, organise live streamed press conferences, build their own legal teams, etc. Dealing with such a diverse range of individuals while policing a demonstration represents significant planning and tactical challenge for policing organisations, who would find it easier to plan policing responses if these groups were more homogenous (e.g. football hooligans) (ibid.). Environmental movements and protests are further amplified by the use of digital communication technologies, with visual and textual material being widely shared on social media platforms (Di Ronco & Allen-Robertson, 2021). A good example of this amplification is how the frequency and perceived impact of these protests can now be traced online.[1]

---

[1] See for example Climate Protest Tracker https://carnegieendowment.org/publications/interactive/climateprotest-tracker#

Protesters actively challenge the system and the status quo, yet they are often criticised for not contributing to it. This is why not all interviewees agreed on the extent and actual impact of these demonstrations. However, Amir Niknam highlighted how highway blockades and the activity of Extinction Rebellion have raised more awareness about climate change within the Dutch Police.

> I think early 2023, the topic started getting a lot of traction again, mainly because of Extinction Rebellion and the highway blockades. I noticed a lot of people within the police started thinking about the climate. (Amir Niknam)

These actions have also helped sustain momentum around discussions of ecocide legislation, which can be classified as an example of so-called positive 'radical flank dynamics', meaning how the presence of radical factions within a social movement influences (positively or negatively) the movement's overall success and public perception (Simpson et al., 2022). This was strongly echoed in a conversation with Jojo Mehta, who cited Polly Higgins, a Scottish environmental activist best known for her work advocating for the recognition of ecocide as an international crime:

> I mean the grassroots mobilisations have been absolutely fundamental in the ecocide movement because they opened up the space where the ecocide law conversation could sit. Ten years ago, ecocide law was considered an extreme thing and people didn't want to support it. After Extinction Rebellion, Fridays for Future and all of that, those movements kicked off suddenly, and we look like common sense in comparison. In fact, Polly Higgins used to say, I love XR - they make us look moderate.

The preparation for policing eco-demonstrations might now involve considerable police resources at all stages of the protest. Some of them are informed by the intelligence-led policing model and might include surveillance of key activists, monitoring their communication strategies, planting under-cover officers amongst activists, and developing a network of informants (Button et al., 2002). There have also been attempts to apply proactive initiatives to liaise with protest that draw on community policing philosophy and are realised in the form of dialogue, mediation, negotiation, sensing the moods within a crowd, sustaining communication throughout the protest, upholding prior agreements (Gorringe & Rosie, 2013; Holgersson & Knutsson, 2010). There are examples of introducing dialogue as a police strategy to respond better to eco-demonstrations. However, this is still unknown territory for the police that is regarded more as pressure valves and competes with the knee-jerk reaction to contain, manage the crowd, and when police mediators or dialogue facilitators are seen as 'traitors' by fellow colleagues and as 'intelligence gatherers' by protesters (Gorringe & Rosie, 2013).

Despite these demonstrations being perceived as more relatable and less harmful, a former police officer from the United Kingdom observed that the police as a whole are still not adequately equipped to manage them effectively:

> [The police] are not geared up or not geared up to deal with very specialist environmental protest. For example, not every police force has the capability to respond to protest at height where people climb up buildings or infrastructure. That causes a problem for them because if people climb up a large building, how do the police deal with it now? Some police forces, larger metropolitan ones or specialist police forces do have rope access, protest intervention teams. That is one of the jobs that I used to do. We used to deploy across Britain, even into

Scotland, to deal with people climbing up things to contain it and to bring them down in many cases. And also in tunnels, where people are tunnelling. Not every police force is equipped to deal with that. (Interviewee 1)

The main issue, and the reason for addressing this topic at this particular point in the book, is not merely the inadequacy of police responses to environmental demonstrations, but rather the broader concerns surrounding the criminalisation of environmental activism and the tendency to label all forms of protest as eco-terrorism. Given that the latter is increasingly entering public discourse, there needs to be clearer operationalisation of what is meant by the 'radical wing' of eco-demonstrations. While it is true that these demonstrations are not monolithic, existing along a spectrum that includes pressure groups, radical environmental pressure groups, militant environmental activist groups, and even terrorist organisations, the increasing use of state police to suppress reformist activism has drawn significant criticism. This includes the use of aggressive crowd-control tactics, unjustified infiltration of activist groups, or surveillance of their social media accounts (Di Ronco & Allen-Robertson, 2021; Goulka et al., 2023; Gulliver et al., 2023; Pellow, 2023).

Furthermore, Goulka and colleagues, based on observations from the United States, argue that criminal justice agencies and mechanisms can also be used to curtail attempts addressing the climate crisis. Changes in the crime/harm landscape might also include the introduction of new types of crimes that aim to curtail environmental activism and protect the interest of corporate organisations (e.g. through the redefinition of what counts as 'critical infrastructure' or the provision of immunity to drivers who hit protesters with their cars enacted, e.g. in Florida, Iowa, and Oklahoma). So-called gag laws are powerful examples of the criminalisation of efforts and attempts to collect information about harmful practices of corporate polluters (Goulka et al., 2023). Another example is the US animal industry and corporate advocacy groups advocating for legislation named 'Agroterrorism Prevention Act' or the 'Animal Enterprise Terrorism Act', to counteract climate/environmental activism, including the sentencing option of a death penalty for anyone who would cause a lethal activist arson (ibid.). This begun to be accompanied by the introduction of additional security and intelligence measures applied by targeted companies. The so-called securification of companies comes in the form of heightened security fencing, security guards, access control systems, security procedures, or the reliance on private intelligence networks and investigators (Button et al., 2002, p. 30). With the advent of more specifically targeted protests, the increasing role of private sector involvement in responding to eco-protestors raises issues concerning their accountability (ibid.).

Insufficient communication, along with inconsistent guidance on managing public disorder, might also lead to radicalisation and resentment not only among the public but also within the police service itself. This issue is rarely acknowledged in the field of criminology, where scholars too often assume that while the streets are dynamic and evolving, the police as an organisation remain static and unchanging. This was highlighted by a Dutch police officer as follows:

I have many colleagues who are absolutely furious about the blocking of the A12 [Dutch motorway] time and time again. Last year we were confronted with the farmers and we were told to wipe these people off the field. Or the people who came protesting against the vaccinations were water blasted. Water cannons and dogs were set loose and everything. It's very hard for us to understand why you would hit so hard on one group and give the others free rein. (Interviewee 3)

The streets are not the aspect we should focus on, as environmental and climate demonstrations, as important as they are, do not represent entire societies. In an interview with Alexandra Jones it was highlighted that it is not only the environmental activism that we see on the streets that matters, but equally is about the quiet segment of society that release their anger, frustration, disappointment, and lack of trust in governments in voting booths:

This too may provoke a sort of radicalisation. It's just not directly physical on the streets, but it is visible in the voting booths, especially among lower working class or lower middle class. Very ambitious climate policies may provoke further disengagement and distrust towards the government. We need these people to go forward because they are the bulk of the nation. You say the squeaky wheel gets the grease? Those are the young people screaming on the A12. They get all the attention. They do not necessarily represent a large share of votes. We've seen that in the last elections [in the Netherlands]. I think it would be very unwise not to pay attention to it. Given the polarisation going on the other side of the spectrum, because people see it as rich peoples' ideas for which they have to pay, and they're not convinced that it does them any good. (Alexandra Jones)

Although responding to civil disobedience or public disorder has been part of policing's repertoire for decades, the task of policing environmental and climate demonstrations reveals vulnerabilities in law enforcement unlike any other consequence of climate change. It draws the police into activities that may become increasingly criminalised, which in turn could further erode the already declining public trust in the police.

For example, the legitimacy of climate-related policies created by governments is closely interlinked and depends on the trust in institutions that enforce these policies. Trust and confidence in the police, and the way the police do policing, significantly influence public attitudes and preferences regarding climate policies. Findings from the European Social Survey (ESS), collected in or around 2016 across 23 European Countries, reveals that people's acceptance and support for policies, such as those promoting renewable energy, primarily rely on trust in impartial institutions, such as the legal system and the police, rather than partial entities such as a government (Kulin & Johansson Sevä, 2021). Furthermore, while climate change overall as an exogenous shock indicated by Roché and Fleming (2025) is indicated as potentially sparing the erosion of public trust in the police, the application of police force during such protests (and then conceptually viewed as an endogenous shock) might bring the opposite effects.

These demonstrations also provoke discussions within police services, which can lead to progressive change, or, in some case, radicalisation due to a lack of clarity. Therefore, the issue of policing environmental and climate demonstrations deserves particular research attention, as it is shaped by multiple and sometimes competing discourses. To date, I have not encountered a criminological study that fully captures this complexity or offers a balanced representation of the perspectives involved. While such demonstrations undoubtedly raise awareness of environmental and climate-related harms, they also entail a significant trade-off: they may strain public trust in law enforcement, challenge the legitimacy of policing practices, and risk contributing to internal polarisation within police forces.

Policing 'Greenlash' Demonstrations

Although increasingly prominent, the topic of policing demonstrations that oppose green or climate change policies has received less criminological attention. As previously mentioned, the climate crisis is closely linked to the use of fossil fuels and CO_2 emissions. Effectively responding to the climate crisis necessitates significant transitions, including the transformation of current unsustainable farming practices. Unlike eco-demonstrations, which aim to show strong support for climate protection policies, 'greenlash' protests are organised to publicly resist policies related to climate and environmental goals, such as the EU Green Deal, decarbonisation, elimination of on-road emissions, or ESG (environmental, social and governance) investment goals, all of which are central to these transition processes.

Similar to environmental and climate demonstrations, these protests also present a challenge for police organisations. However, as noted by Amir Niknam, they may garner more sympathy than environmental protests, as they are not advocating for change but rather for the preservation of the status quo:

> I would say that any institution will find it easier to empathise, collaborate, and perhaps facilitate a movement aimed at maintaining the status quo, rather than one that seeks change. (Amir Niknam)

The cost of the energy transition has already sparked numerous pushbacks from businesses and industries directly impacted across different parts of the world. Especially farmers' protests, which have become increasingly prominent across Europe (CNBC, 2023), have emerged as a symbol of such 'greenlash' reactions. The proposed EU Green Deal has triggered protests in countries including Belgium, France, Germany, Greece, Hungary, Italy, Latvia, Lithuania, the Netherlands, Poland, Spain, and Romania.[2] In Poland, the 2024 farmers' protests aimed to resist the implementation of the EU Green deal, while also expressing resentment over the increased influx of goods from Ukraine, which were made possible by the 'solidarity corridors' introduced in 2022 by the EU Commission.[3] In Poland, these greenlash demonstrations are part of a broader and coordinated effort, led by the legendary

---

[2] https://www.carbonbrief.org/analysis-how-do-the-eu-farmer-protests-relate-to-climate-change/. Accessed 6.03.2024.

[3] https://www.euractiv.com/section/politics/news/polish-farmers-rally-against-green-deal-ukrainian-foodimports/. Accessed 6.03.2024.

Solidarity trade union, which is calling for a national referendum on the matter.[4] Similar protests have occurred in India. Analysis of the farmers' protests in India since the summer of 2020 reveals that public discourse around these protests has been dominated by agrarian distress and grievances, with ecological dimensions rarely acknowledged in the public debate (Baviskar & Levien, 2021).

In the Netherlands, following a landmark court ruling in 2019 that mandated a reduction in excess nitrogen emissions,[5] due in part to the production processes requiring considerable fossil energy, significant protests erupted across the country. The analysis of the 2019 protests, as well as earlier similar demonstrations, shows a complex range of grievances fuelled by anger, disappointment, a sense of disrespect, and perceptions of unjustified limitations on business growth. Given the violent history associated with similar movements,[6] there was even a prospect of army deployment to manage the farmer's protests in 2019 (van der Ploeg, 2021). The successful farmers' lobby during these demonstrations set a precedent for current resistance to policies aimed at mitigating the climate crisis.

The introduction of EU policies aimed at addressing the environmental effects of contemporary farming has been met with varied reactions across EU member states. The discourse and organisation of these protests differ from country to country. In some places, resistance focuses on farming subsidy cuts, rising farming costs and taxes, or EU trade deals with non-EU countries (Dwyer, 2024). All these protests stall the EU's overall progress towards implementing climate-friendly policies and meeting CO_2 emission targets.

According to van der Ploeg, Professor of Economics at the University of Oxford and the Research Director of the Oxford Centre for the Analysis of Resource Rich Economies (OXCARRE), the current state of agricultural landscape in Western Europe presents a worrying picture, as many of these protest groups can be characterised as radical movements that quickly attract support and sympathy from rightwing populist political parties (van der Ploeg, 2021). Ursula von der Leyen's observation that the pesticides proposal 'has become a symbol of polarisation'[7] illustrates the complex struggle to introduce and enforce new, progressive environmental policies amid backlash from certain segments of society.

What are the implications for law enforcement? Both types of demonstrations, those advocating for change and those resisting it, serve as significant yet complex indicators of how the consequences of climate change are increasingly impacting and distressing ordinary citizens. While climate activism has garnered considerable attention within criminology, there is a notable lack of discussion surrounding greenlash protests. Furthermore, as highlighted by one expert, police

---

[4] https://preczzzzielonymladem.pl/. Accessed 5.08.2024.

[5] In the Netherlands, 46% of all nitrogen is emitted by animal husbandry (van der Ploeg, 2021).

[6] For example, in the 1990s there was an emergence of a strongly populist and violent movement called NVV [Nederlands Vakbond van Varkenshouders: Dutch Syndicate of Pig Producers] (van der Ploeg, 2021).

[7] https://www.theguardian.com/commentisfree/2024/feb/16/europe-farmers-climate-green-protest-eu. Accessed 16.02. 2024.

responses and sympathies may vary considerably depending on the nature of the protest to which they are deployed. Additionally, the application of different standards to different protests may contribute to police frustration and influence their behaviour towards protesters. From a criminological perspective, while the presence of right-wing extremism within greenlash demonstrations is increasingly acknowledged, there remains a notable absence of critical debate regarding the potential for left-wing extremism and its associated vulnerabilities within pro-climate protest movements.

Civil disobedience, in whatever form it takes, represents one of the most demanding aspects of the 'climatisation' of police work, shaping not only the way the police operate but also their public image and legitimacy. The irony is that, regardless of the government's political stance, right- or left-wing, some form of protest or civil disobedience will always inevitably persist, requiring continuous police engagement.

References

Baviskar, A., & Levien, M. (2021). Farmers' protests in India: Introduction to the JPS Forum. *The Journal of Peasant Studies, 48*(7), 1341–1355.

Blaustein, J., Miccelli, M., Hendy, R., & Hutton-Burns, K. (2023). Resilience policing and disaster management during Australia's Black Summer bushfire crisis. *International Journal of Disaster Risk Reduction, 95*, 103848.

Button, M., John, T., & Brearly, N. (2002). New challenges in public order policing: The professionalisation of environmental protest and the emergence of the militant environmental activist. *International Journal of the Sociology of Law, 30*(1), 17–32.

CNBC. (2023). From Washington to Warsaw, a 'greenlash' is picking up steam despite extreme heat. Retrieved from https://www.cnbc.com/2023/08/01/extreme-heat-a-greenbacklash-is-sweeping-across-the-us-and-europe.html. Accessed 6.03.2024.

Di Ronco, A., & Allen-Robertson, J. (2021). Representations of environmental protest on the ground and in the cloud: The NOTAP protests in activist practice and social visual media. *Crime, Media, Culture, 17*(3), 375–399.

Dwyer, O. (2024). Analysis: How do the EU farmer protests relate to climate change? Retrieved from https://www.carbonbrief.org/analysis-how-do-the-eu-farmer-protests-relate-toclimate-change/. Accessed 6.03.2024.

Gorringe, H., & Rosie, M. (2013). 'We *will* facilitate your protest': Experiments with liaison policing. *Policing: A Journal of Policy and Practice, 7*(2), 202–209.

Goulka, J., Kang, S., Martin, A., Ensign, K., & Beletsky, L. (2023). When crises collide: Mapping the intersections of climate, pollution, crime and punishment. Northeastern University School of Law Research Paper No. 437, *Northeastern University Law Review,15*(2). Retrieved from https://ssrn.com/abstract=4453693 Accessed 15.02.2025.

Gulliver, R., Banks, R., Fielding, K. S., & Louis, W. R. (2023). The criminalisation of climate change protest. *Contention, 11*(1), 24–54.

Holgersson, S., & Knutsson, J. (2010). Dialogue policing—A means for less collective violence? In T. Madensen & J. Knutsson (Eds.), *Crime prevention studies: Preventing collective violence* (pp.191-216). Willan Publishing.

Kulin, J., & Johansson Sevä, I. (2021). Who do you trust? How trust in partial and impartial government institutions influences climate policy attitudes. *Climate Policy, 21*(1), 33–46.

Martiskainen, M., Axon, S., Sovacool, B. K., Sareen, S., Furszyfer Del Rio, D., & Axon, K. (2020). Contextualizing climate justice activism: Knowledge, emotions, motivations, and actions among climate strikers in six cities. *Global Environmental Change, 65*, 102180. https://doi.org/10.1016/j.gloenvcha.2020.102180

Pellow, D. N. (2023). Confronting institutional violence in the context of climate justice politics. *Social Media + Society, 9*(2). https://doi.org/10.1177/20563051231177913

Reiner, R. (2009). Police property. In A. Wakefield & J. Fleming (Eds.), *The Sage dictionary of policing* (pp. 230–231). Sage.

Roché, S., & Fleming, J. (2025). The underplayed importance of shocks in policing studies. *Policing and Society: An International Journal of Research and Policy, 35*(4), 381–397.

Simpson, B., Willer, R., & Feinberg, M. (2022). Radical flanks of social movements can increase support for moderate factions. *The Proceedings of the National Academy of Sciences Nexus, 1*(3). https://doi.org/10.1093/pnasnexus/pgad478

Van der Ploeg, J. D. (2021). Farmers' upheaval, climate crisis and populism. *The Journal of Peasant Studies, 47*(3), 589–605.

Chapter 6
Policing Disasters

Police as an Emergency Management Actor

The frequency and intensity of disasters have significantly increased over the past decades and this trend is predicted to continue (IPCC, 2022; Thomas & López, 2015; WMO, 2021). Between 2000 and 2020, there were 4623 climate-related disaster events, affecting more than 3.39 billion people (many on more than one occasion), approximately 44% of the global population in 2020, and led to 2.97 trillion in economic losses worldwide. These events resulted in the deaths of over 472,000 people (CRED, 2020; Donatti et al., 2024). The most significant disasters in terms of human impact were droughts, affecting over 1.4 billion people worldwide, followed by riparian floods, the most frequent disaster, which impacted more than 1.2 billion people. Tropical cyclones were the deadliest, causing nearly 190,000 deaths (Donatti et al., 2024).

In a meticulous review of global hotspots of climate-related disasters in 172 countries by Donatti et al. (2024), climate-related disasters are classified as: (1) meteorological disasters, caused by short-lived extreme weather and atmospheric conditions lasting from minutes to days; (2) hydrological disasters, resulting from the occurrence, movement, and distribution of surface freshwater and saltwater; and (3) climatological disasters, driven by long-term atmospheric processes, with variability ranging from intra-seasonal to multi-decadal periods. The review includes disasters such as droughts, tropical cyclones, flash floods, riverine floods, coastal floods, wildfires, heatwaves, landslides, and mudslides.

The 2024 review by Donatti and colleagues demonstrates a clear link between the number of people affected by disasters and the Human Development Index (HDI). Countries with a higher HDI had less people impacted by these events, likely due to their greater capacity to implement adaptation strategies. This is evident, as wealthier nations can afford access to early warning systems and necessary infrastructure. They also tend to have a stronger tradition of drafting and executing evacuation plans, along with greater financial resources to support essential changes. It

© The Author(s) 2025

A. Matczak, *Adapting to Climate Change in Modern Policing*, SpringerBriefs in Criminology, https://doi.org/10.1007/978-3-031-97510-3_6

is important to remember this in line with scholarly perspectives in disaster studies, which argue that disasters often result from human mismanagement, and a lack of effective action in managing hazards. As a result, the concept of 'natural disaster' is considered a misnomer (Kelman, 2020). The true disaster is not the immediate event itself, but rather an accumulation of poorly informed and misguided decisions, or the absence of action, leading up to it (ibid.).

Disasters not only test the limits of human suffering but also challenge the resilience of first responders to human emergencies, a dynamic compellingly examined in the scholarship of Jarrett Blaustein and colleagues. The analysis of police responses in East Gippsland to Australia's Black Summer bushfires (2019–2020) demonstrated how police can effectively form, support, and collaborate within an emergency management network (Blaustein et al., 2023). The case study highlighted how police officers in a remote rural community proactively contributed to disaster management, shaped by their prior experiences, the rural policing environment, and institutional reforms inspired by previous high-profile cases. The authors argue that the police's ability to act as resilient emergency management actors in this context should be seen as an extension of community- and problem-oriented policing, with a strong tradition of building partnerships within the community.

However, the authors also note that police responses to disasters might be negatively affected when emergency management is not considered a police priority or when there is a history of difficult police-public relations. Effective disaster management depends on the region's vulnerability prior to the event and the extent to which the severity of environmental hazards had been previously underestimated. Inappropriate and/or inadequate policing response, can, in fact, exacerbate the human security impacts of the disaster (Mutongwizo et al., 2022), as it was evident in the 2010 Haiti earthquake, or during the Hurricane Katrina. The latter has gained a lot of scholarly attention due to the extent and scope of police failures during a time of crisis (Adams & Stewart, 2014; Deflem & Sutphin, 2009; Kotey, 2009). The case study highlights those police responses in New Orleans were inadequate in maintaining order, requiring the assistance of both public and private actors. The response demonstrated that when police fail to fulfil their duties in a country with a long history of police militarisation, other actors step in, either by mandate, as with federal agencies, or through recruitment, as with private military companies. The inadequacy of the police was further compounded by long-standing distrust among minority communities, exacerbated by pre-disaster reports of discrimination and corruption within the New Orleans police force.

Disasters now occur in a highly globalised world with fragmented social structures, hybrid governance of relationships across multiple sectors and actors, with an increasing role of technology (Sakurai & Murayama, 2019). As a result, public health emergencies and natural disasters create, more than ever, the urgent need for the state to share authority, legitimacy, and capacity with other bodies. Exploring how different police actors respond to a disaster provides a unique perspective to study plural policing during which are marked by an unprecedented weakness in state responses and increasing socio-economic vulnerabilities. Holley et al. (2020) argue that the future of policing and security governance in these areas is confronting a continuing decentralisation of security along with an expansion of private security actors (both human and nonhuman). Therefore, it is quite likely that with the increasing frequency and severity of disasters, there will be an emergence of unique disaster plural policing networks, as there are already evidence that high-consequence, complex disasters accelerate the pluralisation of policing at a time of crisis as policing (security) actors are required to engage and forge multi-agency alliances.

The variety of policing configurations during large-scale crisis situations might intentionally or unintentionally amplify existing securitisation trends. Securitisation processes are often manifested in the presence of the military and/or militarised responses to natural disasters. In case of Hurricane Katrina, this process was additionally amplified by increased commercialisation of policing tasks during the disaster. Humanitarian operations produce distinct spatial formations such as camps and air archipelagos, which may resemble military operations in certain socio-political contexts (Brankamp, 2019). While Adams and Anderson (2019) argue that the presence of the military is sometimes necessary to restore order during a disaster, humanitarian emergencies, including those caused by disasters, require 'a care mindset' (Benton, 2017), which often clashes with the presence and activities of the army at the scene. Although NATO is recognising its added responsibility from not only responding to armed conflicts but also to an increased engagement in responding to natural disasters (Global Affairs Canada, n.d.), the future challenge for practitioners, policymakers, and scholars alike will be to balance the perils and opportunities that come with disaster plural policing in existential crisis situations. At the same time, this may cause political and academic critics of securitisation to lambast plural policing as the central culprit, without considering the potential benefits of such practices in weak or fragile states or critical emergency situations. The proverbial baby may be dispensed with alongside the bathing water (Matczak et al., 2021).

Policing Disaster Crimes and Harms

The role of the police as an emergency management actor, as argued by Blaustein and colleagues, is one of several ways in which the language of climate change is entering police organisations. From the conceptual perspective advanced in this book, it also constitutes a clear manifestation of the climatisation of police work. However, disasters represent a dual role for the police: not only as a key part of a broader emergency management network focused on noncrime activities, but also as law enforcement, where they are expected to maintain their core responsibilities. This chapter serves as a bridge between the discussion of how climate change is reshaping the crime and harm landscape and how these changes are influencing the police as an organisation—a process discussed separately in this book under the chapter 'Green organisational reset'. Such perspective was sparked by a comment made by the environmental law expert from the Netherlands, suggesting that

policing disasters can align with the well-established timeline used in disaster studies. The range and diversity of crimes and harms the police may be called upon to
address that are related to a disaster can fit into the framework long applied in disaster management (Alexander, 2002; Coppola, 2015):

> In disaster management, we say there are three phases. First, there is the ex-ante phase, or
> 'time minus one,' which includes prevention and precaution. Then comes the disaster itself,
> or 'time zero,' when relief measures must be taken. Finally, there is 'time plus one,' which
> focuses on recovery and compensation after the disaster. (Interviewee 9)

The same expert then emphasised that, at the pre-disaster stage, it is crucial for the
police to be actively involved in drafting, implementing, and preparing disaster
management plans. In terms of disaster preparedness and planning, particularly in
the context of climate change, police should play a key role in advising and drafting
disaster management strategies. Their involvement is critical both in the ex-ante
phase (preparation and planning) and at time zero (the moment of disaster), where
their duties include ensuring public safety, protecting people and property, and facilitating emergency shelter.

Following this expert's observation, the obvious question arises: when we work
on disaster preparedness and management plans, do we have the police at the table?
With few exceptions, there is a significant lack of research on police preparedness
for climate-related disasters. However, findings from the 2024 study on the police
and partnership preparedness for the consequences of the climate crisis in the United
Kingdom suggest that when the police are ill-prepared, as they are in the UK context, it is due to several factors: a narrow interpretation of legislation, short-term
planning, insufficient funding and resources, and a limited ability to anticipate
future challenges (Lydon et al., 2024).

Pre-disaster prevention can include various initiatives, many of which have yet to
be explored by criminologists, particularly regarding how they relate to the shifting
dynamics of crimes and harms. Some of these initiatives are cost-effective and
accessible strategies often rooted in nature. As a result, interventions like nature
conservation, restoration, and management can play a crucial role in helping communities adapt, particularly in regions with low human development and those that
have faced severe climate impacts (Donatti et al., 2024). In my conversation with
Clifford Shearing, he said that criminology has massively overlooked the business
of insurances and its impact on the crime and harm landscape, especially before,
and after disasters. This is also something that as highlighted by the environmental
law expert previously quoted in this section:

> I have been working on disaster-related issues for almost 25 years. In 1992, I had only been
> at the university for a year when a major river flooding occurred. The citizens of the affected
> villages came to us—my colleague, who is now an Advocate General at the Supreme Court,
> and me—asking about compensation. We started investigating, and at that time in the
> Netherlands, there was no insurance available for flooding. This surprised us, so we looked
> deeper into the issue. We discovered that after the catastrophic 1953 flooding, which devas-
> tated half of the province of Zeeland and caused numerous casualties, insurers had formed
> a cartel. As a result, they classified earthquakes and flooding as uninsurable risks and pro-
> hibited their members from offering coverage for such disasters. My colleague and I found
> this decision quite cynical. Why was this the case? (Interviewee 9)

While disasters may also lead to disaster-specific crimes, such as scams or identity theft for insurance fraud, there is a growing discussion about the concept of 'survival crimes', which posits that individuals who cannot access public services during disasters may resort to property crimes or other minor infractions to meet their basic needs. Enforcing laws against such crimes can be seen as a violation of basic human rights, and it is argued that such laws should be relaxed in these situations.[1] However, in my interview with Rob White, green criminologist based in Australia, he pointed out that distinguishing between survival and non-survival crimes during disasters can be particularly challenging for police officers:

> It can sometimes be very difficult to distinguish between looting and survival. If someone desperately needs drinking water, they might break into a house to get it, but that's survival, not looting. Context matters greatly in these situations. From an institutional perspective, however, we are not investing enough in funding, staff training, or equipment to manage the crises we are already facing. This lack of preparedness does not bode well for the future.

Among the many unexplored issues for police during disasters is, for example, the intersectionality between gender and disasters. A security expert from India highlighted this by providing various examples, such as how gender affects the vulnerability of disaster-affected individuals, their likelihood of being targeted for human trafficking, and how gender dynamics influence collaboration during disaster response. These are critical areas that require further research, not only in the field of policing but also across broader disaster management and humanitarian fields. Although the quotation below addresses the victimisation of women in the context of disasters, it can also be interpreted in reverse, as a lens through which to examine the vulnerabilities and strengths of female police officers operating in disaster settings.

> There is evidence showing that many girls, for instance, are forced to leave school when their families are affected by floods. Their education is disrupted because they are often required to take care of their families or, in some cases, are married off due to the loss of their parents' livelihoods. Additionally, in certain regions, such as West Bengal, there have been cases of girls and women becoming victims of human trafficking, particularly those directly affected by climate change-related disasters. These issues are still not well studied or fully understood. For example, what specific dynamics make women more vulnerable in these situations? When disasters strike, many women are not trained in essential survival skills like swimming or climbing trees, which significantly reduces their ability to escape danger or recover afterward. However, in many places, women have taken the lead in disaster preparedness and response. Their awareness and proactive efforts have played a crucial role in encouraging families to evacuate ahead of time. (Interviewee 5)

Then, at the post-disaster stage, Rob White observed that one key issue in Australia is the ongoing challenge of addressing post-disaster fraud. For instance, during the Summer Bushfires of 2019–2020, various crowdfunding initiatives, charities, and government helplines were launched to provide financial aid for those affected. However, amid these relief efforts, fraudulent activities emerged. Scammers posed as insurance brokers or other financial intermediaries, convincing victims to hand

---

[1] https://www.city-journal.org/article/crimes-of-survival. Accessed 1.05.2024.

over the funds they had received from the government or charities under the guise of helping them process insurance claims and rebuild their homes. In reality, these were fraudulent schemes.

Taken together, these insights suggest that, to support police institutions in effectively preparing for and responding to disasters, it is beneficial to draw upon established disaster scholarship and apply the well-recognised disaster management cycle as a framework for identifying and mapping potential crimes and harms that may arise in disaster contexts.

The Humanitarian Pull

Disasters act as 'symmetry-shattering events', meaning a disruptive occurrence that breaks the balance or established order of a system that disrupt established processes and create opportunities to tailor emergency responses more effectively to the specific circumstances (Comfort, 1996, p. 137). Previously cited literature and interview quotations suggest that during disasters, the police take on roles beyond law enforcement, often becoming integral to emergency management, but also demonstrating a more humanitarian approach. Traditionally framed within security governance, policing during public health emergencies or disasters shifts towards global or humanitarian governance, moving away from state-centred authority. Unlike security governance, which operates within the realms of politics and power, humanitarian governance is more aligned with ethical considerations (Barnett, 2013). This perspective was echoed by the security expert from India:

> The police will play a crucial role in both emergency preparedness and emergency response, this is already happening. However, they must coordinate effectively with other agencies, as working in isolation will not be effective in the future. One major gap is that police often operate in silos, which is unsustainable, especially when it comes to disaster management. Whether in pre-disaster warnings, response efforts, or post-disaster recovery. I am specifically referring to state police here. Their role must go beyond simply maintaining law and order; it needs to incorporate a more humanitarian perspective. The police must adapt and recognize how disasters impact different communities in different ways. A humanitarian approach is essential in addressing these challenges effectively. (Interviewee 5)

There is considerable evidence that climate change can also stimulate collective action that increases community cohesion and adaptation actions. Disasters can serve as turning points that might unite community members and can be seen as an opportunity for an emergence of shared identity around the need to prevent future disasters (Jetten et al., 2021). Such a humanitarian angle was already explored in the field of policing studies. One of the unintended consequences of the COVID-19 pandemic was to highlight the changing police role while facing a public health emergency. This changing security governance environment has made the police services broaden their mandate from a predominant concern with crime to other sources of insecurities and disorder (see Bartkowiak-Théron et al., 2022; Mutongwizo et al., 2022). Such a police involvement in 'crisis situations' can also be observed in a wide array of other case studies, as is robustly presented in the systematic review by Laufs and Waseem (2020).

Each high-consequence disaster displays unique spatial and temporal dynamics connected to the development of the disaster, its impact on the communities, and unique opportunities to analyse the strengths and vulnerabilities of the police and policing processes during the response to the disaster (Adams & Anderson, 2019). Humanitarian operations are difficult to sustain without the not-so-invisible hand of the police. It has long been noted that law enforcement officers are often the first responders called to a scene and they are expected to have significant involvement in disaster-related activities; however, their efforts are rarely acknowledged (ibid.). What emerges from this discussion is the question when we plan and discuss disaster preparedness and management plans—do we have the police at the table?

References

Adams, T. M., & Anderson, L. R. (2019). *Policing in natural disasters. Stress, resilience, and the challenges of emergency management.* Temple University Press.

Adams, T. M., & Stewart, L. D. (2014). Chaos theory and organizational crisis: A theoretical analysis of the challenges faced by the New Orleans police department during Hurricane Katrina. *Public Organization Review, 15,* 415–431.

Alexander, D. (2002). *Principles of emergency planning and management.* Oxford University Press.

Barnett, M. N. (2013). Humanitarian governance. *Annual Review of Political Science, 16,* 379–398.

Bartkowiak-Théron, I., Clover, J., Martin, D., Southby, R. F., & Crofts, N. (Eds.). (2022). *Law enforcement and public health: Partners for community safety and well-being.* Springer.

Benton, A. (2017). Whose security? Militarization and securitization during West Africa's Ebola outbreak. In M. Hofman & S. Au (Eds.), *The politics of fear: Médecins sans Frontières and the West African Ebola epidemic* (pp. 25–50). Oxford University Press.

Blaustein, J., Miccelli, M., Hendy, R., & Hutton-Burns, K. (2023). Resilience policing and disaster management during Australia's Black Summer bushfire crisis. *International Journal of Disaster Risk Reduction, 95,* 103810.

Brankamp, H. (2019). 'Occupied enclave': Policing and the underbelly of humanitarian governance in Kakuma refugee camp, Kenya. *Political Geography, 71,* 67–77.

Centre for Research on the Epidemiology of Disasters (CRED). (2020). Human cost of disasters (2000–2019). Retrieved from https://www.cred Accessed 10.07.2023.

Comfort, L. (1996). *Self organization in disaster response: The Great Hanshin, Japan earthquake of January 17, 1995.* National Hazard's Center, University of Colorado. Retrieved from https://hazards.colorado.edu/uploads/basicpage/QR78.pdf Accessed 20.05.2024

Coppola, D. P. (2015). *Introduction to international disaster management* (3rd ed.). Butterworth-Heinemann.

Deflem, M., & Sutphin, S. (2009). Policing Katrina: Managing law enforcement in New Orleans. *Policing, 3*(1), 41–49.

Donatti, C. I., Nicholas, K., Fedele, G., Delforge, D., Speybroeck, N., Moraga, P., Blatter, J., Below, R., & Zvoleff, A. (2024). Global hotspots of climate-related disasters. *International Journal of Disaster Risk Reduction, 108,* 104488. https://doi.org/10.1016/j.ijdrr.2024.104488

Global Affairs Canada. (n.d.). NATO Climate Change and Security Centre of Excellence. Retrieved from https://www.international.gc.ca/world-monde/international_relationsrelations_internationales/nato-otan/centre-excellence.aspx?lang=eng. Accessed 10.09.2024.

Holley, C., Mutongwizo, T., & Shearing, C. (2020). Conceptualising policing and security: New harmscapes, the anthropocene, and technology. *Annual Review of Criminology, 3,* 341–358.

IPCC. (2022). In H.-O. Pörtner, D. C. Roberts, M. Tignor, E. S. Poloczanska, K. Mintenbeck, A. Alegría, M. Craig, S. Langsdorf, S. Löschke, V. Möller, A. Okem, & B. Rama (Eds.), *Climate change 2022: Impacts, adaptation, and vulnerability. Contribution of Working Group II to the sixth assessment report of the intergovernmental panel on climate change.* Cambridge University Press. Retrieved from https://report.ipcc.ch/ar6/wg2/IPCC_AR6_WGII_FullReport.pdf Accessed 26.07.2023

Jetten, J., Fielding, K. S., Crimston, C. R., Mols, F., & Haslam, S. A. (2021). Responding to climate change disaster. *European Psychologist, 26*(3), 161–171.

Kelman, I. (2020). *Disaster by choice: How our actions turn natural hazards into catastrophes.* Oxford University Press.

Kotey, P. (2009). Judging under disaster: The effect of hurricane Katrina on the criminal justice system. In J. Levitt & M. Whitaker (Eds.), *Hurricane Katrina: America's unnatural disaster* (pp. 105-131). Lincoln.

Laufs, J., & Waseem, Z. (2020). Policing in pandemics: A systematic review and best practices for police response to COVID-19. *International Journal of Disaster Risk Reduction, 51,* 101812.

Lydon, D., Hallenberg, K., & Kapageorgiadou, V. (2024). 'This is not a drill': Police and partnership preparedness for consequences of the climate crisis. *International Journal of Police Science and Management., 27,* 16. https://doi.org/10.1177/14613557241248295

Matczak, A., Voss, K., & Ashmore, D. (2021, January 14). Plural policing during health emergencies and natural disasters. *Leiden Security and Global Affairs Blog.* Retrieved from https://www.leidensecurityandglobalaffairs.nl/articles/plural-policing-during-healthemergencies-and-natural-disasters. Accessed 11.09.2024.

Mutongwizo, T., Blaustein, J., & Shearing, C. D. (2022). Resilience policing and climate change: Adaptive responses to hydrological emergencies. *CrimRxiv.* https://doi.org/10.21428/cb6ab371.fb7dfcd0

Sakurai, M., & Murayama, Y. (2019). Information technologies and disaster management–benefits and issues. *Progress in Disaster Science, 2,* 100012. https://doi.org/10.1016/j.pdisas.2019.100012

Thomas, V., & López, R. (2015). *Global increase in climate-related disasters.* Asian Development Bank. ADB Economics Working Paper Series, No. 466. Retrieved from https://www.adb.org/sites/default/files/publication/176899/ewp-466.pdf Accessed 10.07.2023.

WMO. (2021). *Weather-related disasters increase over past 50 years, causing more damage but fewer deaths.* World Meteorological Organisation. Retrieved from https://wmo.int/media/news/weather-related-disasters-increase-over-past-50-years-causing-more-damage-fewer-deaths Accessed 10.07.2024.

Chapter 7
Green Organisational Reset

Climate change spares no country, institution, or sector from its effects. Although its impact is distributed unevenly, it will reach everyone sooner or later. Much like the rise of artificial intelligence, climate change compels every institution and sector to engage with it, take a position, and answer the question: What does the impact of climate change mean for the organisation you work for? How will it transform it? The term coined in this book 'green organisational reset' refers to an organisational, and potentially cultural, shift within police services reflected in changes to police operations, practices, strategies that align with principles of sustainable and environmental responsibility. Within some police services, a discernible process of 'green organisational reset' is underway, some of which are discussed in connection with the impact of climate change. This chapter examines two dimensions of this transition: (1) the implementation of sustainability strategies and the greening police operations, (2) the expansion of police mandates through the establishment of specialised environmental crime units.

Sustainable Operations

Although the integration of climate change considerations into police organisations is painfully slow, it is happening through different, often diverge, channels that shape their actions and responses to some degree. In Australia, for instance, this occurs through the language of emergency management (see Blaustein et al., 2023), while in other countries, it is driven by environmental protests or policy requirements aimed at making police operations more sustainable. The latter considers initiatives such as green mobility, fleet electrification, circular waste economy, eco-friendly police buildings, and/or promotion of sustainable practices in general.

© The Author(s) 2025
A. Matczak, *Adapting to Climate Change in Modern Policing*, SpringerBriefs in Criminology, https://doi.org/10.1007/978-3-031-97510-3_7

The introduction of sustainability policies can now be observed in individual countries like the England and Wales (APCC, 2022), the Netherlands (Sybrandij, 2023), Finland (Poliisi, 2021), Canada (Konyk, 2018), Philippines (Romualdo & Robles, 2023), or as a focus of collaborative projects within larger entities such as the EU's ENLETS (European Network of Law Enforcement Technology Services) Green Policing Procurement initiative (ENLETS, 2023). As these processes are still in their early stages, they remain subject to experimentation and potential flaws. During my interview with a senior EUROPOL employee (Interviewee 11), as well as during the CERIS workshop (Community for European Research and Innovation for Security) on the impact of climate change on security practitioners,[1] it was expressed that the introduction and sustainable policies might face significant implementation obstacles due to various concerns. Among these concerns are worries about reduced police efficiency and a lack of market response to accommodate specific police design requirements (e.g. to green police fleets or dedicated police charging stations to avoid cybersecurity issues). Similar concerns have long been observed in the military sector as well. However, a senior security expert from the Netherlands suggested that it would be helpful to view these challenges as opportunities for police organisations:

> [Sustainable operations] show the opportunities that are in climate change as well. There are huge opportunities if you can create military units that are fully self-sustaining. (Interviewee 12)

Self-sustaining and environmentally sustainable police operations are particularly crucial in countries and disaster-affected areas where stable power grids are unavailable or unreliable. In the scholarly literature this is called 'energy autarky', referring to a state in which a region, country, or organisation achieves complete energy self-sufficiency, meaning it generates all the energy it needs without relying on external sources. This concept is often associated with renewable energy and decentralised energy systems, allowing entities to become independent from national or global energy markets (Müller et al., 2011).

While the introduction of sustainability language and operations is a highly appreciated development, David Lydon pointed out that it still represents a rather narrow understanding of the impact of climate change on police work.

> They [National Police Chief Councils, England] did come back to us, and interestingly, they thought that we were talking about how green the police are. For example, the electrification of their vehicle fleet. They thought that's what it was about. Normally there are two things that you get from senior police officers. One, it's about electrifying car fleets and reducing their carbon footprint. The second one is dealing with environmental protest. There was no other kind of wider vision about the role of policing in contingency planning for the wider consequences of climate change.

Although it might initially be seen as having a limited perspective, the language of sustainability can serve as a buy-in—a path of least resistance—to establish a

---

[1] https://home-affairs.ec.europa.eu/news/ceris-workshop-impact-climate-change-security-practitioners-202405-06_en

longer-term narrative about the impact of climate change on police organisations and their work. A senior security expert from the Netherlands commented on this as follows:

> You address it from an operational perspective, because that's what gives the acceptance in the organisation. If I force everybody to do more about climate change, you get a lot of resistance. That's very important. What kind of storyline do you choose for the police? How do you bring about the climate change message?
>
> Do you bring it as an operational message? (Interviewee 12)

While there is a scarcity of research on climate change communication within police context, there is much to be gained from the extensive body of research on climate change communication and education (see Moser, 2016). Equally important is the need to understand both supporters and sceptics, including outright deniers of climate change, while recognising that individuals do not make decisions in isolation. A network perspective highlights that individuals are embedded within dynamic and interconnected social relationships, the nature and structure of which can significantly influence the transmission of attitudes and behaviours (Leombruni, 2015).

Studies co-created with practitioners on police sustainability indicators in the Philippines demonstrate that the narrow understanding of sustainability, primarily in terms of greening police practices, can evolve to encompass broader environmental aspects. This includes topics such as green behaviour and awareness, environmental law enforcement, or community partnerships in pursuit of pro-environmental actions (Romualdo, 2022; Romualdo & Robles, 2023). At the time of writing, a similar development is occurring within the Dutch Police, where the initial sustainability project is expanding in scope and being integrated into the broader environmental safety policing agenda. This process likely reflects a broader narrative of making the police a more sustainable organisation while simultaneously undergoing structural and operational transformation.

The remaining question is of implementation and the actual efficiency of such sustainable policies, and whether the police really understand the concept of environmental impact, as most police organisations fail to examine the environmental impact of their operations (Konyk, 2018).

This was also emphasised in my interview with David Lydon:

> The police in the UK announced in the last year that its ambition is to be the most science and technology led police service in the world. A lot of that is through partnership with private industry. That's problematic because then you are the police that is going to end up protecting the interests of big corporations and big tech. My specific question is about the environmental impact of technology. Do the UK police understand this type of environmental impacts on a global scale when embracing science and technology?

At the outset of this book, I argued that the climatisation of police work and police as an organisation involves the distribution of 'hurt' and trade-offs. The transition to sustainable police operations has the potential to spark and expand debates on the impact of climate change within the organisation. This shift is often closely tied to an increased reliance on technology, or policy recommendations to adopt such technologies, which are themselves not entirely free from environmental harm. Training a single AI model can result in a significant environmental impact, emitting as much carbon as five cars in their lifetimes (Strubell et al., 2020). While the transition to sustainability within police organisations is inevitable, there is a pressing need for a deeper understanding for their true carbon footprint and the complexities of their environmental impact.

Environmental Crime Enforcement

The second avenue through which police services undergo a green organisational transition is through capacity-building processes for environmental crime enforcement and investigation, which is increasingly recognised as a critical issue (Colantoni et al., 2022; UNODC, 2024). As previously established in the book, although not all environmental crimes are directly linked to climate change, both climate change and the broader sustainability transitions inherently involve environmental considerations. The rapid expansion of environmental legislation and criminalisation necessitates robust monitoring, supervision, and enforcement, often requiring collaboration across diverse and sometimes distant sectors. Unlike most traditional crimes, environmental harms demand significantly greater resources, specialised expertise, and enhanced interagency information sharing and collaboration.

Policing environmental crimes has undergone several evolutionary stages, beginning with the first police responses in the 1950s, the emergence of initial regulatory and administrative powers in the 1970s, and the introduction of market-based instruments in the 1980s and 1990s (Holley & Shearing, 2016). This was followed by the launch of specialist environmental protection agencies, environmental courts, and new fractions of police (green or conservation police). The complexity of policing, steering, and addressing environmental harms has evolved in tandem with the increasing recognition of these harms as well as their increasing complexity and transboundary nature, for example, more polluters, more complex, transnational, and multilayered environmental crimes (White, 2011; Holley & Shearing, 2016). The interdependence between the complexity of enforcement and nature of environmental crimes was also pointed out by a senior environmental law expert from the Netherlands:

> These kinds of mechanisms, biodiversity, nature protection, it becomes much more difficult to say exactly what the norms and the standards are. And so there I would say that that enforcement also is much more complicated. (Interviewee 9)

National institutional frameworks for environmental crimes typically comprise complex networks of bodies and agencies that can generally be categorised into those with law enforcement responsibilities and those with administrative roles. Law enforcement authorities, including national police forces, are part of a broader assemblage of actors that also includes customs agencies, maritime and coastal authorities, border control agencies, tax authorities, and organisations responsible for managing forests and national parks. These entities all contribute significantly to enforcement efforts. On the other hand, regulatory and administrative bodies are tasked with a variety of responsibilities, including monitoring, licensing, conducting pre-trial proceedings, and initiating investigations (Colantoni et al., 2022).

Most recently, the assemblage of policing and regulatory actors responsible for responding to environmental crimes and harms has been analysed in criminology within the framework of new environmental governance (NEG). NEG is a continuing social experiment that shows promise in terms of public-private partnership, flexibility and adaptation, information exchange, open-ended standards, reactive and proactive mediating role, all of which are essential to addressing the dynamic and multifaceted causes of environmental harms (Holley & Shearing, 2016). There is now a widespread agreement that successful environmental crime enforcement requires horizontal and vertical cooperation between states, relevant agencies, communities, and international bodies, which represents the ongoing shift towards network-based and pluralised forms of environmental security and governance (ibid.), which is still argued in the scholarly literature as something that can substantially improve the effectiveness and efficiency of responses to environmental harms. This is particularly important due to increasing transnational nature of environmental crime (Gore et al., 2019). The regulation, policing, and enforcement of environmental harms will continue to be a shared responsibility between administrative and criminal justice agencies. However, recent developments in the EU, especially the implementation of the revised EU Environmental Crime Directive, suggest that this framework may undergo significant changes in the future, moving away from its overreliance on administrative mechanisms towards enhanced protection and stricter punishment for environmental violations (Colantoni et al., 2022). This evolution, of course, implies an increased role and involvement of police agencies in identifying and investigating environmental harms.

Emergence of Environmental/Green Police Units and Partnerships

The necessary multi-agency approach to combating environmental crimes has not impeded police efforts to build capacity for environmental crime enforcement and investigation. In recent years, significant progress has been made in this area, with many examples of police organisations establishing or expanding specialised units and actively participating in various networks and projects to share expertise and enhance collaboration.

At the EU level, since 2017, there have been systematic, concerted efforts to build a strategic approach towards environmental crime enforcement. This includes the launch of the '4network' of judicial (EUFJE—European Union Forum of Judges for the Environment) and prosecutorial professionals (ENPE—European Network of Prosecutors for the Environment), administrative and inspectorate bodies (IMPEL—European Union Network for the Implementation and Enforcement of

Environmental Law), and police and other enforcement forces (EnviCrimeNet) (Colantoni et al., 2022). In addition, the Intelligence Project on Environmental Crime (IPEC), the implementation of the European Union Action Plan against Wildlife Trafficking, lobbying for the prioritisation of environmental crimes and the revision of the EU Environmental CrimeDirective contributed to the gradual process of advocating for green policing (Moreno, 2023). The two-year AMBITUS project, led by the French Ministry of the Interior, sought to support EU member states in enhancing their long-term efforts to combat environmental crime. The project provided a comprehensive analysis of environmental crimes, examining them from multiple perspectives, including types of offenses, emerging trends and estimated impacts, enforcement structures, key stakeholders involved, and the challenges hindering effective action (see Colantoni et al., 2022). More recently, the EU-funded projects Emeritus[2] and Perivallon,[3] which include police partners within their consortia, aim to enhance police capacities to address environmental crimes by promoting the more frequent use of advanced technologies. Last but not least, the revised EU Environmental Crime Directive underscores the need for a stronger commitment and more effective capacity-building by police agencies for the enforcement of environmental crime. Outside of the EU, for example, the emergence of green policing within the Philippine National Police was argued as aligning with the core value of *Makakalikasan* (pro-environment), which focuses on environmental protection and conservation to maintain ecological balance (Romualdo, 2022). The emergence of these specialised units is likely to serve as a precursor to, or catalyst for, the development of dedicated climate change-related policing teams in the future.

Is this enough? While the growth of environmental legislation, and the subsequent promise of environmental crime enforcement in many countries, is satisfactory to some extent, it is the implementation and readiness to act on it that is a problem, as observed by Jojo Mehta and critical criminologist from the United Kingdom:

> One of the biggest problems we've had with environmental law, which has grown up over the last few decades in a really prolific way, is that there are many bodies of law in different countries. We have a fundamental problem with people taking it seriously. That is a cultural issue. That's a very deeply embedded cultural issue. We are the lords and masters of nature. We take from it as we wish. (Jojo Mehta)

The proliferation of environmental law policies and legislation is not followed by a serious commitment on the practical, operational, and enforcement levels:

> If we were really serious about dealing with environmental crime, then we would dedicate the police force, I don't know, say 50% of police resources to enforce the environmental crime. Now, that would change something. That would be a cultural change. It's almost impossible. (Interviewee 15)

---

[2] https://emeritusproject.eu/

[3] https://perivallon-he.eu/introduction-to-perivallon-project-2/

Despite growing attention from policymakers and enforcement agencies, criticism persists regarding the inadequacies in addressing environmental crimes. These include an inadequate, or poorly integrated legislative framework, insufficient operational tools, lenient penalties, limited specialisation among criminal justice actors, inadequate data availability, weak information sharing, and insufficient international cooperation (Colantoni et al., 2022). In addition, whichever developments have occurred these were possible despite dwindling police resources, when environmental law enforcement often being the first to suffer. Sharif and Uddin (2021), while analysing the Bangladeshi context, talk about the 'lax practice' nature of the law enforcement mechanisms, which they consider to be most of the time reactive and merely for show. There is a lack of regular monitoring, remedial actions, follow-up, collaboration, and preventive strategies. And this occurs despite the fact that, Bangladesh is a signatory to around 25 international conventions, protocols, and agreements on environmental protection and is prolific in terms of domestic regulatory and legal documentation, with at least 200 rules, regulations, and policy papers related to the environment. This is perhaps, historically speaking, environmental crimes are still a 'novelty' in the crime and harms landscape.

Shrinking or inadequate police resources are frequently cited as the biggest obstacle to building the capacity needed for successful environmental crime enforcement (Interpol, 2022; Ozymy & Ozymy, 2022). The limited resources and lack of commitment were the main reasons why the revised EU Directive, adopted on 11 April 2024, significantly strengthens police powers by requiring the developing national strategies, increased resources, specialised training, and enhanced cooperation mechanisms within and between EU member states (European Parliament and Council, 2024). Although some may question the promise and the actual level of implementation of the Directive, it is the anticipation of law that is already shaping the reality.

There is no comprehensive international legal framework that fully protects the environment or universally defines and criminalises environmentally harmful behaviours. Instead, environmental protection is governed by a scattered system of international and regional agreements, inconsistently incorporated into national legal systems. This lack of cohesion results in regulatory gaps and enforcement weaknesses, creating opportunities for criminal and economic interests to exploit the system and infiltrate legitimate industries (UNODC, 2024). However, as Jojo Mehta strongly emphasised: 'What's being underestimated is the impact of knowing it's coming, because that's actually what starts to shift behaviour and starts to shift the thinking'. The same expert interviewee also highlighted the shaming aspect of criminal law, something that potentially contradicts the premise of the new environmental governance described earlier.

> The only thing that actually can potentially curb that [environmental crimes] at a fundamental level is criminal law, not regulatory law, because regulation, there's no moral stigma attached to breaching regulation. You might get a slap on the wrist or a fine. So people tend to sort of budget for that or work out how to not fit into certain categories that are forbidden and that sort of thing. As soon as you put a criminal law in the mix, you have a very different, approach, a very different sense of the importance. And that obviously is a really fundamental thing at the cultural level. (Jojo Mehta)

The infrastructure for environmental crime enforcement, as well as the potential to foster an eco-cop mindset, understood in this book as pro-environmental attitude that is interwoven with a commitment to promote and execute policing tasks related to protecting the environment, is significantly influenced not only by the social, political, and economic circumstances of the country, but also by the changing environment itself. For instance, Christian Bueger's analysis (2020) of environmental crime enforcement in Kiribati, Tuvalu, and Palau highlights how severe weather events impact police infrastructure. In particular Kiribati and Tuvalu have extremely limited sea-going capacity and number of police staff available. New patterns of blue crimes, the growing influx of tourism, and increasing concerns about biodiversity and the import of alien invasive species already make it impossible to execute efficient police responses in these areas.

Lastly, policing and investigating environmental crimes will increasingly necessitate the use of advanced technologies and the application of digital and online tools for evidence collection, as highlighted by many of my interviewees. The tools can range from satellite imagery, phone tracking devices and installations, surveillance, or artificial intelligence. However, this approach presents a certain sustainability trade-off, as the deployment of technology itself contributes to environmental harm—an issue that was already discussed in this chapter.

> I think it will broaden the range of what will need to be looked at in court. And, you know, that may well involve digital evidence, it may well involve satellite data, it may well involve all of those kinds of things. (Jojo Mehta)

The uncritical and unreflective use of technology has been shown to lead to unintended social consequences, such as social sorting, discrimination, increased public distrust, and so-called function creep. The latter describes a phenomenon driven by an unquestioned reliance on technological solutions and the expansionary nature of surveillance tools, resulting in their application in contexts or for purposes beyond their original intent (van Brakel & de Hert, 2011).

Policing Sustainable Transition Processes

In my conversation with the British critical criminologist, a significant emphasis on the need for increased police involvement in the investigation of environmental crimes was highlighted, particularly those involving white-collar and corporate offenses. The expert pointed out that: 'we are in for a time of more intense competition' and projected that despite the growing regulatory frameworks and the criminalisation of such activities, these forms of harm are likely to persist within the context of the ongoing climate emergency.

Simultaneously at the same time as police services will have to increasingly reshape their operations as sustainable, they might be obliged to increasingly monitor and police the sustainable transition processes of others. Recent years have seen a shift away from enforcement and deterrence to compliance and non-prosecution agreements. At this stage these processes are monitored and regulated by private actors but it is possible to imagine that this might well be a police task in the future, as pointed out by one of the environmental law experts from the Netherlands:

> The tendency that we have now is that some of these industries are presenting themselves as "we are green". You don't have to tell us anything, we will do compliance and control. We will control our whole supply chain because we have an interest. We will make sustainability reports, and so on. This, of course, from a policy perspective, also from a policing perspective, also raises the question how should you view that? On the one hand, we can say absolutely great. However, the capacity of the police is limited. They cannot do everything. (Interviewee 9)

> With all of these internal compliance mechanisms, the tendency in policing is that the police and the inspectors no longer control the real pollution or the real emissions, but they control whether there is compliance with the internal compliance mechanism. What are they doing now? They look at the sustainability reports and they say whether the sustainability report looks nice. Of course, sustainability reports can look beautiful. You can pay an accountant enormous money to make a wonderful sustainability report that says nothing on your real emissions. (Interviewee 9)

Similar findings emerged from the AMBITUS project cited earlier. For instance, the expanding regulations introduced under the EU Green Deal initiative are anticipated to create new opportunities for crimes and violations, such as VAT fraud linked to the expansion of the EU Emissions Trading System and the illegal trade of banned chemicals under the updated Registration, Evaluation, Authorisation, and Restriction of Chemicals (REACH) regulation (Colantoni et al., 2022).

Climate change, air pollution, and biodiversity loss are deeply interconnected. While climatisation of police work is giving impetus to the capacity-building process for environmental crime enforcement, while preparing the book, I occasionally encountered conversations where environmental crimes and their enforcement were considered in isolation from the consequences of climate change. This may be because we have not yet built a sufficient and sustainable infrastructure within the police to effectively address environmental crimes, making it difficult to see the bigger picture that includes the impact of climate change and adds to the overall complexity.

However, at the practice level, there are examples where the lines between environmental crime enforcement and green or sustainable policing initiatives are becoming blurred and merged. Scotland's 2021 'Environmental Strategy', which follows the country's pioneering proclamation of a climate emergency in 2019, is a document that outlines efforts to make police operations and organisation more sustainable (Police Scotland, 2021). In the Netherlands, several initiatives were launched as a joint effort under the umbrella term of environmental security between the Sustainability Unit and Environmental Crime Strategic Advisory to raise awareness and create a sense of urgency around both environmental crime investigations and the impact of climate change. The 2022 Interpol Background Paper 'Crime, Law Enforcement and Climate Change' views the domain of environmental crimes as naturally aligned with the implications of climate change. Given the significant scarcity of expertise and resources within law enforcement to address environmental issues, the convergence of efforts on both environmental and climate-related matters may become an increasingly common trend in the future.

References

Association of Police and Crime Commissioners. (2022). *PCCs making a difference*. Environment and sustainability in focus. association of police and crime commissioners. Retrieved from https://www.apccs.police.uk/media/7867/apcc-environmental-and-sustainability-in-focus.pdf Accessed 10.10.2022.

Blaustein, J., Miccelli, M., Hendy, R., & Hutton-Burns, K. (2023). Resilience policing and disaster management during Australia's black summer bushfire crisis. *International Journal of Disaster Risk Reduction, 95*, 103810.

Bueger, C. (2020). Rising waters. *The impact of climate change on maritime law enforcement agencies in Pacific Island Countries. Report for the Global Maritime Crime Programme of UNODC*. Retrieved from https://www.researchgate.net/publication/370902753_Rising_Waters_The_Impact_of_Climate_Change_on_Maritime_Law_Enforcement_Agencies_in_Pacific_Island_Countries Accessed 26.03.2024.

Colantoni, L., Sarno, G. S., & Bianchi, M. (2022). *Fighting environmental crime in Europe: An assessment of trends, players, and actions*. Ambitus Project Report. Istituto Affari Internazionali (IAI). Retrieved from https://www.iai.it/en/pubblicazioni/c09/fighting-environmental-crime-europe-assessment-trends-players-and-action Accessed 10.08.2024.

Directive (EU) 2024/1203 of the European Parliament and of the Council of 11 April 2024 on the protection of the environment through criminal law and replacing Directives 2008/99/EC and 2009/123/EC. Retrieved from https://eur-lex.europa.eu/legalcontent/EN/TXT/?uri=CELEX%3A32024L1203 Accessed 1.09.2024.

ENLETS. (2023). *Moving towards a greener, sustainable future for policing in Europe*. Retrieved from https://enlets.eu/moving-towards-a-greener-sustainable-future-for-policing-ineurope/ Accessed 12.08.2024.

Gore, M. L., Braszak, P., Brown, J., et al. (2019). Transnational environmental crime threatens sustainable development. *Nature Sustainability, 2*(9), 784–786.

Holley, C., & Shearing, C. (2016). Policing and new environmental governance. In B. Bradford, I. Loader, B. Jauregui, & J. Steinberg (Eds.), *The SAGE handbook of global policing* (pp. 552–572). SAGE.

Interpol Innovation Centre. (2022). *Crime, law enforcement and climate change*. Interpol Innovation Centre. Background Paper. Retrieved from file:///C:/Users/annaf/Downloads/IC%20Background%20Paper%20-%20Crime%20Law%20Enforcement%20and%20Climate%20Change-3.pdf Accessed 16.08.2024.

Konyk, J. (2018). *Green policing: Recommended actions for an environmental sustainability plan for the vancouver police department*. Retrieved from https://sustain.ubc.ca/sites/default/files/2018-54%20Green%20Policing%20Recommended%20Actions%20for%20an%20Environmental%20Sustainability%20Plan%20at%20the%20VPD_Konyk.pdf Accessed 16.08.2024.

Leombruni, L. V. (2015). How you talk about climate change matters: A communication network perspective on epistemic skepticism and belief strength. *Global Environmental Change, 35*, 148–161.

Alfaro Moreno, J. A. (2023). Towards an understanding of consequences of the prioritisation of environmental crime in the European Union. *Cuadernos de la Guardia Civil, 69*, 31–48.

Moser, S. C. (2016). Reflections on climate change communication research and practice in the second decade of the 21st century: What more is there to say? *WIREs Climate Change, 7*(3), 345–369.

Müller, M. O., Stämpfli, A., Dold, U., & Hammer, T. (2011). Energy autarky: A conceptual framework for sustainable regional development. *Energy Policy, 39*(10), 5800–5810.

Ozymy, J., & Ozymy, M. J. (2022). The green police: Criminal enforcement in the era of climate change. *Environmental Law Reporter, 52*(7). Retrieved from https://www.elr.info/articles/elr-articles/green-police-criminal-enforcement-era-climate-change Accessed 17.08.2024.

Police Scotland. (2021). *Environmental strategy.* Retrieved from https://www.scotland.police.uk/spa-media/uzhblrhq/environmental-strategy.pdf Accessed 1.09.2024.

Poliisi, Police of Finland. (2021). *Keeping everyone safe, at all times. Finish Sustainability Report 2021.* Retrieved from https://poliisi.fi/documents/25235045/28127375/Poliisinvastuullisuus raportti-2021-en.pdf/ac6de219-6c26-16b5-93c1-30169c93eae3/Poliisinvastuullisuusraportti-2021-en.pdf?t=1665576793438 Accessed 12.08.2024.

Romualdo, A. (2022). Prospect for green policing: Constructs and dimensions of environmental sustainability in the context of public safety. *Global Sustainability Research, 1*(2), 22–29.

Romualdo, A., & Robles, A. C. M. O. (2023). Towards a sustainable green policing: A Delphi-based forecast of sustainability indicators for law enforcers. *Global Sustainability Research, 2,* 52. https://doi.org/10.56556/gssr.v2i3.556

Sharif, S. M., & Uddin, M. K. (2021). Environmental crimes and green criminology in Bangladesh. *Criminology & Criminal Justice, 23*(3), 490–510.

Strubell, E., Ganesh, A., & McCallum, A. (2020). Energy and policy considerations for modern deep learning research. *Proceedings of the AAAI Conference on Artificial Intelligence, 34*(09), 13693–13696. https://doi.org/10.1609/aaai.v34i09.7123

Sybrandij, I. (2023). Duurzaamheid is voor de politie meer dan alleen invulling geven aan het Klimaatakkoord. *Cahiers Politiestudies, 68,* 45–58.

United Nations Office on Drugs and Crime (UNODC). (2024). *Global analysis on crimes that affect the environment—Part 1: The landscape of criminalization.* United Nations Publication.

White, R. (2011). Environmental law enforcement: The importance of global networks and collaborative practices. *Australasian Policing: A Journal of Professional Practice and Research, 3*(1), 12–16.

Van Brakel, R., & De Hert, P. (2011). Policing, surveillance and law in a pre-crime society: Understanding the unintended consequences of technology-based strategies. *Cahiers Politiestudies, 3*(20), 163–192.

Chapter 8
Resilience and Adaptive Policing

In criminological and policing scholarship, the concept of resilience has garnered significant attention, particularly in preparing police services for the challenges posed by climate change. This chapter contributes to this body of work by highlighting a crucial indicator of organisational resilience: the role of the police as a central node in expanding partnerships, assemblages, and multi-agency collaborations that necessitate the pluralisation of policing responsibilities. Additionally, it draws on psychological literature, which posits that a resilient mindset entails openness to learning and adaptation. If we accept the former, we must also embrace the latter.

The recognition of the Anthropocene has given rise to two distinct approaches to addressing climate change consequences. One line calls for a more top-down, state-orchestrated approach, and the other proposes the opposite, which is decentralisation of security/policing tasks (Harrington & Shearing, 2017; Holley et al., 2020). The third avenue for police organisations is to explore and implement greening efforts, that is, the adoption of resilience and/or adaptive policing models. The emergent model of 'resilience policing' has been proposed in this field of study as a promising solution that builds upon established state-based community policing traditions to respond to increasingly complex risks caused by disasters (Mutongwizo et al., 2022; Blaustein et al., 2023). The idea of resilience policing goes beyond the resilience of individual police officers and police organisational responses to occupational stress and extends the established forms of community policing, where police officers act as facilitators and enablers who assist communities to effectively withstand the adversities of disasters (ibid). Resilience policing involves a significant shift in the established mentalities of the police, community, and other security actors to appreciate the 'care for the collective good mindset', which encourages a wide variety of nodes to embrace ways of working together as self-organised and polycentric learning networks (see Harrington & Shearing, 2017).

However, from the practical perspective, a closer evaluation of, for example, the UK-based Local Resilience Forums (Lydon et al., 2024; Malik & Berg, 2024),

A. Matczak, *Adapting to Climate Change in Modern Policing*, SpringerBriefs in
Criminology, https://doi.org/10.1007/978-3-031-97510-3_8

which are considered the key local and regional multi-agency collaborations to respond to civil emergencies, including those induced by climate change, reveals very little connection to the real realisation of resilience as proposed in the literature. The Local Resilience Forums fall short of fulfilling their intended purpose, representing a missed opportunity for effective implementation, which is meaningful collaborations that harness diverse exchange of knowledge and experience in order to tackle the multilayered manifestations of climate change impacts (Malik & Berg, 2024). Instead, these ill-prepared units are continually hampered by a narrow focus on official and formal requirements, short-term planning, tradition of 'silos' working, lack of funding and resources, as well as limited prescience (Lydon et al., 2024).

Blaustein et al. (2023) point out that it remains unclear whether a resilience policing model should be conceptualised as a gradual extension of existing community-oriented or partnership-based approaches to complex harms, or as a novel development in security governance that may transform police and policing in an age of ecological crises and catastrophes. The concept of resilience was received in the field with mixed responses (see Blaustein et al., 2024), and similar criticism was raised by a couple of my interviewees, who saw it a techno-managerial concept, proposed as something not being able to operationalise:

> Resilience is an overexploited concept. It depends on how you want to go about it. If resilience is all about communities and individuals being left on their own to build their own resilience, that's not going to work. If structural inequities, post-colonial experiences, gender identities, none of these are taken into consideration, then resilience is a very techno-, scientific-techno-managerial concept. It can lead to more problems than actually solving them. (Interviewee 5)

Alexandra Jones also noticed that it is evident that those discussing resilience consider themselves secure and often hold positions or exist in environments that provide extensive support. However, they may be unaware that, without these external safeguards, they would struggle to sustain themselves independently.

Just like with all other policing models and philosophies, the concept of resilience policing is most likely to remain limited to specific social, cultural, political, and economic contexts, where resilience policing would make sense in countries where there is a tradition of community policing and where the police are seen as representatives of communities.

Resilience is usually seen as an ecological term and some question its transferability to social science (Olsson et al., 2015). Hence, in an interview with David Lydon he argued that climate change and policing needs to be reframed as a community safety matter (rather than emergency management and planning) as it would emphasise better what is already known about police work, including the protective role of the police and their caring mindset towards the people they serve:

> I think going forward, this needs to be about reframing a lot of the consequences of climate change as community safety, local community safety issues. Whereas at the moment the role of the police is very much around emergency planning and preparation.

Due to the inherent limitations of climate predictions, the new security measures emphasise strengthening people's adaptive abilities, particularly through social capital, and bolstering their resilience to the current changes in weather patterns (Oels, 2012). Although adaptation in the climate change jargon is associated with defeatism, a more realistic and promising direction and call to action comes from the recent proposal of Blaustein et al. (2024) who still draw on the resilience literature but claim that it can be a useful starting point for imagining and implementing adaptive policing responses to climate crisis. Adaptive policing can be interpreted as a strategy between absorption of shocks and stressors of climate change impacts and a potential transformation. Adaptive policing might also lead to the police embracement of the 'ethos of care' (Harrington & Shearing, 2017), buying humanity additional time, and it may even support the development of more sophisticated approaches to governing vulnerability and police reform (Blaustein et al., 2024). This call for adaptive policing was also echoed in my interview with Alexandra Jones:

> There are no solutions, there are only trade-offs and adaptations. That's what I made of it. It's constantly about making choices, about bad options. There are no good options. It's a political decision, and our politicians typically advocate linear interventions. Which option do I go for? What do I have to do? It's easier for dictators in times of crisis. You need to take into account the severity and probability of frequency of a crisis. When the police are confronted with this new security situation you have to choose between: do we prepare for shocks or do we prepare for stressors? Both require resilience. Can I take the shock or how do I deal with a stressor over a prolonged period of time?

A critical moment during which the resilience and adaptability of emergency services, including the police, can be observed and tested is so-called Day Zero. The term became widely known during Cape Town's, South Africa, water crisis in 2018 and is now commonly used to describe a critical point in time when resources or services (typically water, but also energy, food) are expected to run out or become unavailable unless significant measures are taken to mitigate the situation (Pascale et al., 2020).

This is connected to another interesting observation from the interview with Amir Niknam, which highlights the need to harness and facilitate adaptive skills, which cannot be effectively achieved through ready-made training packages filled with information on the topic.

> We should be prepared for something. However, something like training is too rigid, I would say, we should train for adaptability. That could mean all kinds of things.

While resilience policing refers to the capacity of police organisations to withstand, recover from, adapt, and learn from crisis or disruptions, the adaptive model of policing concerns the police capability to adjust their strategies, structures, processes, and tactics in response to the adversity. Although the boundaries between resilience and adaptive policing are somewhat blurred, the ability to adapt to rapidly changing environments is undoubtedly beneficial for both. Educational studies suggest this as a potential future direction in various forms. The growing demand for nonroutine, high-level cognitive tasks and advancements in technology will require more than just the accumulation of new knowledge. Instead, it will necessitate continuous learning, along with the willingness and ability to unlearn, learn, and relearn (Ra et al., 2019:27).

Police as a Collaborative Node

An insightful indicator of police resilience worth exploring in greater depth is their capacity to collaborate, establish, and maintain partnerships, as well as their decision-making role within these constellations. One of the key findings and takeaway from the UK study on the preparedness of Local Resilience Forums (Lydon et al., 2024) to address the impact of climate change was the urgent need for collaborative preparedness. Although some planning and exercise occurred around flooding and drought, among the most significant shortcomings are the tradition and ethos of silo working—when the police work in isolation. The ability to act quickly, to collaborate and communicate efficiently with others during a crisis can be considered one of the key indicators of resilience and adaptability. However, the police's ability to do so has been questioned by one of the police officers from the Netherlands who said: 'It's typically for the police, that everybody does their own thing. They seem to run their own show' (Interviewee 3).

In contemporary systems of public safety and security, policing no longer resides solely with the public police. Instead, the practice of policing has become increasingly nodal, fragmented, and multi-tiered. This process and its outcomes have long been discussed in criminology under the term 'plural policing' explained earlier in the book. This concept refers to the involvement of the police as one among many public, commercial, and voluntary actors in maintaining order, preventing crime, investigating crime, and co-producing public safety.

It is anticipated that the complexity and variety of climate change-induced consequences will further amplify the pluralisation of policing, increasing the involvement of various actors in delivering policing functions. Environmental crime response agencies often organise their activities and partnerships around networks defined by geography (such as key transit points and destinations), specific disciplines (like environmental regulatory bodies), or particular commodities (such as waste). Collaboration within and across these networks can take various forms, including horizontal, vertical, cross-functional, and cross-hierarchical interactions (White, 2022). The benefit of collaborative preparedness, particularly when involving experts in the field and academia, allows for different, but often complementary, perspectives on climate-related problems. This issue is also articulately examined by Malik and Bergh in their analysis of the limitations of policing knowledge within the UK Local Resilience Forums in the context of climate crisis preparedness (Malik & Berg, 2024). While the police will remain an important node within this configuration, the key question is which policing tasks they will continue to steer and which tasks they will be drawn into performing—without leading—in collaboration with other actors in this evolving assemblage. This is a critical process to observe, as emphasised by Clifford Shearing:

> We used to speak about partnerships, police partnerships. This puts the police at the centre and with them choosing who to partner with. That's what community policing was. It was the police getting communities to give them the information they need so they can do their job. It's very police centred. (…) I think the word collaboration is different because it's police that may be enrolled by other people. They are not simply in control of these enrolments. Health systems can enrol them. Emergency systems can enrol them. This notion of collaborative arrangements within policing assemblages where the police are not the controlling node, they're part of an assemblage. Sometimes they take the lead, sometimes they're the followers. This is what we drew attention to in the early work of nodal governance. What is the police position in an assemblage? Are they doing steering or rowing?

Although the emerging literature on disaster and resilience policing acknowledges that police are part of wider networks (see Adams & Anderson, 2019; Laufs & Waseem, 2020; Blaustein et al., 2023), the focus is still predominantly on the activities of national police services. However, Clifford Shearing raises another important question regarding the impact of climate change and the police's ability to collaborate: who are, or will be, the new actors?

> A much bigger part of the policing landscape is private security. Considerably bigger, both in numbers and in budgets, and in scope, actually. One could take the same kind of question about the police, and ask about private security. There hasn't been as much work done on that. Another question, which is one that I've been exploring, is on the emergence of new policing actors. And who are they?

In many of my expert interviews, discussions touched on the role of established and emerging private carbon credit trading verifiers, as well as the criminology field's relative neglect of the insurance sector and its influence in shaping the policing landscape. A closer examination of recent events already begins to offer potential answers to Clifford Shearing's questions.

The research on the impact of heatwaves on the Indian traffic police has been a helpful case study to observe two findings. The first one is on how environmental factors, compared to general population, contribute to the heightened occupational heat stress associated with adverse health outcomes (Raval et al., 2018), which has made the climate change consequences to be seen more as safety and occupational matters. Existing policies are often inadequate to support adaptive police capacity or to mitigate the negative impacts of heat on the Indian traffic police, which leads to the second observation. In order to address the police shortages and the growing demands on the traffic police, the government decided to introduce a new parapolicing actor, traffic brigade,[1] which was noted in my interview with an epidemiologist and public health scholar from India:

---

[1] https://english.gujaratsamachar.com/news/gujarat/gujarat-grapples-with-traffic-police-shortage-trafficbrigade-overworked. Accessed 14.08.2024.

> Another thing which we are noticing now [in India], is that the government, whether to save money or because of difficult finances, is recruiting for the traffic police so-called support personnel. They are not police people, but they are called traffic regulation brigade or something like that. They have a different uniform. They are not government civil servants. They are not selected through the government procedures, but they are like a para police. They are employed by the government, but on a contract. (Interviewee 13)

Another example of a non-state actor taking a policing/investigative role to confront the impact of climate change, but in a different capacity, is the work of the Centre for Climate Crime Analysis (CCCA),[2] a non-profit organisation based in the Netherlands and founded by prosecutors and law enforcement professionals. The objective of the organisation is to apply modern investigative techniques and use information and communication technological tools to produce actionable, user-friendly casefiles ready for relevant authorities to start a proceeding in case of any illegal activities associated with the emission of significant amounts of greenhouse gases. Not being restricted by jurisdictional limitations or any other procedural requirements, the choice of the cases is dictated by the available resources but mostly by the impact a case can have on limiting the problem globally. The workings of the CCCA are informed by the nudge theory—a behavioural concept that suggests small, subtle changes in the way choices are presented can influence people's decision (Thaler & Sunstein, 2008). CCCA's work is enhanced by the application of advanced technology and open-source data, and executed by multidisciplinary teams and a worldwide network of contacts consisting of motivated citizens, other civil society organisations involved in similar work, and other relevant experts and institutions on the ground, whose knowledge might benefit the investigation in localised circumstances. The founders of the organisation describe their approach as National Law Enforcement Support Mechanism and classify it as the support and bottom-up model that offers a response to often inadequate and limited national enforcement capacities (Gallmetzer & Cross, 2022). The concept and development of the CCCA serve as an example of an entity within a polycentric governance framework, meaning a system where multiple, independent decision-making centres operate at different levels but are interconnected, as discussed by Malik and Berg (2024). CCCA openly acknowledges the limitations of policing knowledge, and draws on a range of diverse expertise and skills to address this gap. The need for such diverse expertise and skills in multi-agency collaborations and partnerships, and the appreciation when it happens, was broadly acknowledged in all interviews, too.

Another example of an emerging actor that might play a crucial role in environmental law enforcement are the lay people and civic monitoring as a practice entailing the collective mobilisation of ordinary people who are involved in finding evidence of environmental violations. Inspired by the landmark 2019 court decision issued in Texas, known as the Formosa ruling,[3] and based on the 'Sensing for

---

[2] https://climatecrimeanalysis.org/. Accessed 14.08.2024.

[3] The court found Formosa, a Taiwanese petrochemical company, guilty of violating the US Clean Water Act. This case, brought forward by a civic group, relied heavily on evidence gathered by citizens, including volunteer observations of plastic contamination being discharged into the water over an extended period (Berti-Suman, 2024).

Justice' project findings, Anna Berti Suman (2024) provides a compelling case for civic monitoring as a constructive approach in finding evidence of environmental harms but also mediating environmental conflicts. Civic monitoring appears to be on the rise due to increasing availability of audio and video-recording devices, but growing environmental public awareness. In the evaluation of the impact of climate change on maritime law enforcement capacity in Pacific Island countries, one of the findings was that policing maritime violations by the police is significantly dependent on community reporting in that region (Bueger, 2020).

Berti Suman argues that embracing civic monitoring by the police and other criminal justice actors can be an opportunity for authorities to make more inclusive and responsive governance models. The promising outcome of the project has inspired her to explore quickly developing field of civic-gathered evidence to demonstrate impacts associated with climate change and to promote climate justice. Civic monitoring of climate change impacts identifies not only environmental wrongdoings but also the ripple effect of climate change in human and nonhuman systems conditioned to the various social, political, economic, and cultural circumstances. The other advantage of this type of civic monitoring is the discovery of local adaptation strategies and inclusion of traditionally marginalised populations in decision-making processes (Berti Suman, 2024). In support of this, she cites the 2021 court ruling banning gas flaring in the Ecuadorian Amazon, which was largely informed by a citizen science project that produced spatial information thanks to the involvement of Indigenous and farmer communities. Colantoni et al. (2022) highlighted the siloed approach and lack of communication among actors involved in environmental crime enforcement in their comprehensive report on environmental crimes in the EU. They noted that this issue is particularly pronounced when it comes to civil society, which has the potential to make significant contributions but is often constrained by restrictive laws or a lack of trust from institutional stakeholders.

Furthermore, within the recent years, there have been a number of international, horizontal, initiatives of collaboration aimed at connecting different police organisations to enhance their awareness and knowledge base about environmental crimes and the impact of climate change.

The Environmental Network for Environmental Crime (EnviCrimeNet) started as an informal network in 2011 to be tasked in 2013 by the EU's Standing Committee on Operational Cooperation on Internal Security (COSI) to provide a mapping and scoping exercise about the landscape of environmental crimes in the European Union member states (EnviCrimeNet, 2015). It has continued its activities as a network that offers online communication about most recent developments in the field, training, and advice upon request to relevant authorities. The International Initiative of Law Enforcement for Climate (I2LEC),[4] launched in 2023 by the Ministry of Interior of the United Arab Emirates in partnership with UNODC sets an ambitious

---

[4] https://www.unodc.org/documents/commissions/CCPCJ/CCPCJ_Sessions/CCPCJ_32/Statements/Item_8_UAE.pdf. Accessed 14.08.2024.

plan to unite the global police community and advance their understanding about the implications of climate change on their work.

All of the aforementioned initiatives, whether formal or informal, are undoubtedly positive developments. However, several interviewees, among them a senior security expert from the Netherlands, emphasised the importance of less formal networks and collaborations, which they identified as equally valuable in addressing the impacts of climate change. These types of initiatives have the unique ability to harness human potential and energy both within and beyond organisations. They are characterised by flexibility, fluidity, and freedom from bureaucracy and political agendas.

> My network is a non-political network. That was the best move I made. We have experts from more than 40 countries but they don't represent the country. They are there on a personal title because they are personally concerned about this topic. (Interviewee 12)

In an interview with Amir Niknam, it was suggested that the informal potential within the police, referring to individual officers or smaller networks of them, holds significant promise. This potential allows for more critical reflection on police work and fosters stronger connections with the outside world.

> I think instead of thinking of police within police and outside of police, I think it's better to think in terms of the formal organisation and the informal organisation, because the informal organisation, I think does a great job of reflecting and also does a great job of connecting with people outside of the organisation. Formal organisation I would say is more robust, which is good. Everything has a drawback. The drawback of being robust is that it's difficult to change and it's difficult to reflect and change your ways. (Amir Niknam)

Nonetheless, the topic of collaboration, whether involving the police or any other institution, was also acknowledged to have its drawbacks. In my conversation with Rob White, he pointed out two main pitfalls of, for example, international collaboration. The international domain is quite often permeated with 'gesture talk, gesture politics, and gesture diplomacy', which refer to token and symbolic actions or statements used to signal intentions, gain public sympathy, or show solidarity without committing to concrete, significant change. Furthermore, even if there is a pledge to assist the police in adapting to the challenges of the climate crisis, the national conditions and individual police environments in each country might not support it.

> Most of the discussion of collaboration, networking and resource sharing is at the international level between nation states and agencies located within specific countries. We can talk all these nice words about collaboration, we can point out at a global level the deficiencies in staff, resources, equipment, and why we need countries to share with each other the technologies, expertise, the equipment and so on. But when you drill down to the national level, the picture gets incredibly complicated very quickly. (Rob White)

At the time of writing this book, there are still relatively few initiatives solely dedicated to how police can respond to climate crisis. However, as the consequences become more visible and acute, one might expect an avalanche of actions aimed at cooperation and joint effort to tackle the issue. In other expert interviews, practical concerns were expressed about the quality and extent of collaborative police work. Too many initiatives, especially when started on superficial grounds, can lead to distraction and side-tracking from the actual work.

One way of facilitating meaningful, complementary, and efficient collaborative work is the application of technological tools. As pointed out by Amir Niknam collaboration is a widely shared goal, yet it remains challenging to achieve in practice. Simply having the intention to work together is not enough; effective collaboration requires the right tools and systems. New ways of working are essential, methods that enable seamless information sharing without the need for constant meetings or emails. Currently, governmental collaboration is constrained by traditional communication methods, which limit interactions to known contacts. Without pre-existing connections, collaboration becomes nearly impossible. Additionally, there is a natural ceiling to the number of meetings one can attend, restricting the number of stakeholders involved. To overcome these limitations, digital platforms can play a crucial role by leveraging automated systems and data-driven optimisation to connect the right people at the right time, enhancing efficiency and expanding collaborative networks.

Towards a Green Chain of Collaboration in the Criminal Justice System

Police resilience encompasses not only horizontal cooperation, involving practitioners from administrative and regulatory entities, but also hierarchical, involving prosecutors and judges. While the previous section addressed the need for intersectoral and international police collaborations, this section focuses on intrasystemic coordination between the police and other actors within the criminal justice system. The environmental law expert from the Netherlands particularly emphasised the importance of creating a green criminal justice chain of collaboration through training and practice, ensuring that everyone involved is on the same page.

> The region of Ghent in Belgium has specialised police, specialised prosecutors, specialised environmental enforcers, specialised judges, top judges, member of the UN. Even at the level of the Court of Appeals. Specialized chambers. As a policeman, when I have my file here, I bring it to a prosecutor. He knows environmental law. He's going to prosecute. He brings it to a judge, a "green judge" who knows environmental law. (Interviewee 9)

Investigating and prosecuting environmental crimes, as well as handling climate litigation, are still relatively new practices in the criminal justice chain. It is not only police officers who require environmental and climate change education; prosecutors, judges, and other criminal justice actors, equally need such a training and upskilling. Extensive and relevant information sharing can improve the effectiveness and accountability of governance systems, particularly by facilitating the exchange of knowledge and experiences among local groups (horizontally) and between local groups and broader agencies (vertically) (Harrington & Shearing, 2017). Traditional environmental policing approaches tend to be especially effective when supported by specialist environmental courts or judicial officers with expertise in environmental issues.

Their involvement increases the likelihood of prosecuting offenders and applying appropriate sanctions (White, 2013).

In the context of climate change and rising temperatures, teamwork takes on new significance. This is illustrated by the emerging research on the impact of heatwaves on decision-making processes. Apart from the evidence confirming the negative impact of heat on people's propensity for conflict and potential crime rates, exposure to high temperatures has psychological and cognitive repercussions. The analysis of over 10 million arrests made in Texas from 2010 to 2017, along with additional information on subsequent prosecutions and trial proceedings, suggests that heatwaves can lead to poorer decision-making processes by the police and harsher sentences imposed by judges (Behrer & Bolotnyy, 2024). The researchers suggested that teamwork and pre-emptively adjusting staffing levels, such as sending more police officers to respond to an event on hot days, could be one of the factors helping to alleviate the effects of heat on police decision-making.

Although existing literature and research frequently emphasises that the police are insular and have limited experience working with other organisations efficiently, some of my conversations with police experts suggest otherwise. Contrary to this common perception, many within the police acknowledge that this insular approach represents an outdated paradigm, one that is increasingly unsustainable in the face of accelerating climate change and evolving environmental challenges. This was emphasised during my interview with a senior environmental crime expert from EUROPOL (Interviewee 11) who said that: 'This [police isolation] is the old paradigm. This whole paradigm is a system failure. There are still some people working like this, but they are like old dinosaurs, they are going to be extinguished. The new paradigm is information exchange and cooperation'.

Are Police Services Learning Organisations?

So far in this chapter, I have examined examples and referenced literature suggesting that the police are, or have the potential to be, on a trajectory towards transforming their organisational culture to become more environmentally and climate sensitive, responsible, and proactive. Such changes could substantially enhance their capacity to respond effectively to the consequences of climate change. The concept of resilience and adaptive policing implies the necessity for a growth mindset within police organisations, one that embraces challenges and learns from criticism (see Dweck, 2006).

However, Blaustein et al. (2024) acknowledge that this perspective contrasts with the broader consensus in the literature, which often portrays the police as among the least adaptive, innovative, engaging, or responsive institutions. How do the police learn then? Decades of police reform and varying attempts to improve might suggest so, but even within the organisation, the police are often considered to be slow learners, as highlighted by the police officer from the Netherlands:

> I regret to say that, if we [the police] learn, it's on an individual basis. As an organisation, we are woefully slow in accepting the very idea that we could actually learn something from anybody from the private sector, or from healthcare, or from the military. Police learn with great difficulty. (Interviewee 3)

On closer inspection, however, the reality appears more nuanced. Scholarly literature in this field highlights that policing occurs within a distinct and complex organisational learning context, marked by numerous paradoxes (Tops et al., 2012; Tomkins & Bristow, 2024). Police learning is shaped by seemingly contradictory discourses, such as the tension between fostering organisational learning as a means of enabling both individual and institutional growth, and the increasing demand for rigid, evidence-based, and universally standardised practices. For instance, Tomkins and Bristow (2024) examine two specific paradoxes—codification vs. discretion and transparency vs. occlusion—and their implications for organisational learning in policing, providing an in-depth analysis of these dynamics. The authors argue that learning in police setting is not only about gaining knowledge but also about coping and managing emotions, contradictions, expectations, handling blame, and navigating power dynamics.

While 'greening' efforts within the police are much appreciated, there is now greater necessity to integrate climate concerns into policing strategies. Part of this challenging learning process is the police's dependence on governmental and ministerial decisions, directions, which are either absent or confusing, a point that was repeatedly emphasised in some of the interviews.

Alexandra Jones said:

> We serve as we are an instrument. We are an instrument for other organisations to achieve their goals. If you want the police to prepare, we would first have to know what is it they [the government] want us to be able to do. What do you want the police to be able to do in five years' time? In ten years' time? In 20 years' time.
>
> How does the minister think this will look like? How does she see the police in there? Because if I don't know what she wants, I don't know what I have to do to prepare for it. If she says, we're going to have an entirely different organisation to evacuate people in case of flooding, I don't have to worry about flooding so much anymore.
>
> But if she says no, that's also going to be a police job, I have to worry about flooding. How much are we going to ask of our citizens to prepare for it? The most important thing, and that is a huge issue in this country [The Netherlands], which has no strategic culture.

Forecasting, whether applied to the immediate task of predicting police activities based on weather data or to the more complex endeavour of anticipating future climate change-related crimes, serves as a critical learning tool for enhancing strategic thinking. However, the UK research on police preparedness for climate emergencies identified a number of deficits not only at a strategic but also tactical, and operational level. The UK study participants indicated that the habit of short-term planning and lack of foresight vision, limited funding and resources, infrastructural inadequacies, and silo effects hamper efficient planning for climate change-related risks (Lydon et al., 2024). This was further elaborated on by Alexandra Jones in conjunction with climate stressors and shocks:

> The police are specially geared to dealing with shocks. We're very good with that. We deal with incidents and it's very difficult to get the police to think in terms of stressors sooner, which is what is required. When we think about climate change and its impact, we have to think about long term stresses in society and how it will impact us. And we can't. But we can choose to deal with matters on an incident basis. It's for the politicians and for the government to decide what do we want the police to do. (Alexandra Jones)

The COVID-19 pandemic provided a recent, global learning opportunity for police services worldwide in responding to a public health crisis. It served as a moment for law enforcement to 'rethink their relationship with the state and its human representatives' (Tomkins & Bristow, 2024, p. 34). The pandemic also underscored the interconnectedness of the world, offering valuable insights that can be applied to addressing the challenges posed by climate change.

> It is getting better. It has been getting better in the past few years. Covid has been very helpful with that because now we know that things that happened in China in the past year … it needs to be an awareness on all levels, that things that happen abroad have a consequence. (Alexandra Jones)

However, it also highlighted the complex and often challenging intersection between policing and public health.

> I think it does need some reflection. Particularly where there is this big concern and a big risk that because the police are state actors, what are they in order to pursue this green reset? If we want to call it that way. It may involve them operating in a much more authoritarian way to actually discharge the will of government. It's a very fine line to tread because it could force the police into a position of being authoritarian now. And we've seen a flavour of that during Covid. (David Lydon)

van Dijk et al. (2022) argue that the COVID-19 pandemic has brought the relationship between policing and public health to the forefront, highlighting the many social functions of law enforcement. In the absence of immediate medical solutions, police were tasked with monitoring and influencing individual and collective behaviours through simple daily ordinances, such as spatial restrictions and the mandatory wearing of face masks. Public health thus once again became a central focus of operational policing, as law enforcement agencies took on the responsibility of enforcing public health interventions. These experiences provide valuable lessons in preparation for the more acute and accelerating consequences of climate change.

In addition to the lessons learned from policing during the COVID-19 pandemic, the increasing frequency of climate change and environmental demonstrations raises questions about police neutrality. The manner in which police respond to civil disobedience, particularly in countries experiencing intensifying environmental protests, offers significant opportunities for reflection and self-assessment within law enforcement agencies. Although the long tradition of public protests makes them a constant feature (and stressor) of police work, police organisations still often perceive them as unique shocks. In the Netherlands, the activities of Extinction Rebellion and the motorway blockages were among the factors that sparked debate within the police about their neutrality. Ethical dilemmas faced by the police due to climate crisis were also discussed with David Lydon:

> The police saw themselves as in a way ethically neutral. They were saying that they're there to enforce the law. They're not there to make decisions about the relative right or wrong

from an ethical perspective of what the law is. That was quite an interesting one because we were trying to push the agenda that actually the organisation needs to reflect, because once it gets to the tipping point … is it enough for the police to say, well, we don't have an ethical stance on this because there might be a tipping point where they say, well, we actually do need to. This is going to sound weird, but in a way, we need to be more sympathetic to public opinion. For the police that can be difficult because they see themselves as then somehow turning the police, turning on the state itself, if they side with the population rather than what's legally defined.

Discussions about police neutrality have already begun within the Dutch Police, where an informal 'Green' network of officers was established in 2018/2019. This network led to the creation and signing of a petition advocating for more proactive action on sustainability and climate change. One of the key outcomes of this informal movement was a contribution to a change to the police pension funds, which are now prohibited from being invested in fossil fuel companies. Another significant outcome was the recognition of police officers' right to participate in protests outside of their working hours.

We [the Dutch Police] contemplate what does it mean if police officers want to protest, so that we also made up some rules upon that. It's allowed to protest as long as you do it in your free time and not representing the police. I have been to pretty much every single climate protest in the last few years. (Amir Niknam)

Lastly, what we as academics often overlook is that the police are not a monolithic organisation; they are composed of diverse units with distinct roles and responsibilities. Their learning process is also influenced by how educational opportunities are presented. Effectively communicating or delivering 'the climate change message' must account for this diversity. One of my interviewees, a renowned expert and tireless advocate for raising awareness about climate change (Interviewee 12), mentioned that what works best is delivering an impact-driven narrative about the consequences of climate change, tailored for a specific audience. Climate change has almost become a buzzword, so it needs to be filled with concrete examples and case studies, communicated in accessible language that avoids scientific jargon and instead tells a personal story how climate change affects each of us individually. Such an approach will, at the early stage, require different stories for the traffic police, for police strategic advisors, and for those already engaged in countering environmental crimes. The same expert stated that most people do not read IPCC reports, and scientists often express frustration when their warnings go unheeded. However, the issue is not a lack of concern but often a lack of understanding. Bridging this gap requires demonstrating tangible, real-world impacts and engaging experts from fields such as economics, security, and law enforcement to elucidate the relevance of climate change within their respective domains. When individuals recognise its direct implications for their lives, it becomes 'personal' and they are more likely to engage and respond.

While this book seeks to advocate for a more progressive approach to addressing climate change, Rob White cautioned in an interview that progressive police services may face a challenging predicament; they are often subject to criticism regardless of their actions. Many well-intentioned officers are likely to experience growing frustration as global trends unfold, including the rise of misinformation, climate denialism, and the increasing prevalence of authoritarian governance.

Ironically, the more intense the effects of climate change become, the greater the opportunity for the police to accelerate their learning resilience and develop more adaptive pathways. There are several 'low hanging fruits', meaning actions that the police can already consider doing internally or improve upon, such as raising awareness, joining multi-agency collaborations, and revisiting their training and education curricula to enhance the knowledge base about environmental and climate issues. As long as these actions are not merely 'gesture talk', as discussed earlier, they could contribute to meaningful police preparedness to respond to cascading effects of the climate crisis. Otherwise, there is a serious risk that the police will find themselves on the wrong side of history, becoming more authoritarian and losing the legitimacy of the people they serve.

References

Adams, T. M., & Anderson, L. R. (2019). *Policing in natural disasters. Stress, resilience, and the challenges of emergency management.* Temple University Press.

Behrer, A. P., & Bolotnyy, V. (2024). Heat and law enforcement. *The Proceedings of the National Academy of Sciences Nexus, 3*(5), 1–8.

Berti-Suman, A. (2024). *Civic monitoring for environmental law enforcement.* Edward Elgar.

Blaustein, J., Miccelli, M., Hendy, R., & Hutton-Burns, K. (2023). Resilience policing and disaster management during Australia's black summer bushfire crisis. *International Journal of Disaster Risk Reduction, 95,* 103848.

Blaustein, J., Shearing, C., & Miccelli, M. (2024). Adaptive policing for a climate crisis. *Policing and Society: An International Journal of Research and Policy, 2024,* 1–13. https://doi.org/1 0.1080/10439463.2024.2362713

Bueger, C. (2020). *Rising waters. The impact of climate change on maritime law enforcement agencies in Pacific Island countries. Report for the Global Maritime Crime Programme of UNODC.* Retrieved from https://www.researchgate.net/publication/370902753_Rising_ Waters_The_Impact_of_Climate_Change_on_Maritime_Law_Enforcement_Agencies_in_ Pacific_Island_Countries. Accessed 26.03.2024.

Colantoni, L., Sarno, G. S., & Bianchi, M. (2022). *Fighting environmental crime in Europe: An assessment of trends, players, and actions. Ambitus Project Report.* Istituto Affari Internazionali (IAI). Retrieved from https://www.iai.it/en/pubblicazioni/c09/fighting-environmental-crime-europe-assessment-trends-players-and-action. Accessed 10.08.2024.

Dweck, C. S. (2006). *Mindset: The new psychology of success.* Random House.

EnviCrimeNet. (2015). *EnviCrimeNet: Intelligence project on environmental crime. Report on environmental crime in Europe.* Retrieved from https://www.europol.europa.eu/cms/sites/default/ files/documents/ipec_report_on_environmental_crime_in_europe.pdf. Accessed 14.08.2024.

Gallmetzer, R., & Cross, M.E. (2022). *Unlocking National law Enforcement to effectively address global challenges.* Centre for Climate Crime Analysis. Retrieved from https://climatecrime-analysis.org/wpcontent/uploads/2022/07/law_enforcement_support_model__rg-mec_.pdf. Accessed 14.08.2024.

Harrington, C., & Shearing, C. (2017). *Security in the Anthropocene. Reflections on safety and care.* Bielefeld: Transcript.

Holley, C., Mutongwizo, T., & Shearing, C. (2020). Conceptualising policing and security: New Harmscapes, the anthropocene, and technology. *Annual Review of Criminology, 3,* 341–358.

Laufs, J., & Waseem, Z. (2020). Policing in pandemics: A systematic review and best practices for police response to COVID-19. *International Journal of Disaster Risk Reduction, 51,* 101812.

Lydon, D., Hallenberg, K., & Kapageorgiadou, V. (2024). 'This is not a drill': Police and part-nership preparedness for consequences of the climate crisis. *International Journal of Police Science and Management, 27*(1), 16–30.

Malik, A., & Berg, J. (2024). Polycentric governance, epistocracy and the limits of policing knowl-edge in preparing for the climate crisis. *Policing and Society: An International Journal of Research and Policy*, 1–15. https://doi.org/10.1080/10439463.2024.2389961

Mutongwizo, T., Blaustein, J., & Shearing, C. D. (2022). *Resilience policing and climate change: Adaptive responses to hydrological emergencies*. CrimRxiv. https://doi.org/10.21428/cb6ab371.fb7dfcd0

Oels, A. (2012). From 'securitization' of climate change to 'climatization' of the security field: Comparing three theoretical perspectives. In J. Scheffran, M. Brzoska, H. G. Brauch, P. M. Link, & J. Schilling (Eds.), *Climate change, human security and violent conflict* (pp. 185–205). Springer.

Olsson, L., Jerneck, A., Thoren, H., Persson, J., & O'Byrne, D. (2015). Why resilience is unappeal-ing to social science: Theoretical and empirical investigations of the scientific use of resilience. *Science Advances, 1*(4), e1400217. https://doi.org/10.1126/sciadv.1400217

Pascale, S., Kapnick, S. B., Delworth, T. L., & Cooke, W. F. (2020). Increasing risk of another Cape Town "day zero" drought in the 21st century. *The Proceedings of the National Academy of Sciences, 117*(47), 29495–29503.

Ra, S., Shrestha, U., Khatiwada, S., Yoon, S. W., & Kwon, K. (2019). The rise of technology and impact on skills. *International Journal of Training Research, 17*(sup1), 26–40.

Raval, A., Dutta, P., Tiwari, A., Ganguly, P. S., Sathish, L. M., Mavalankar, D., & Hess, J. (2018). Effects of occupational heat exposure on traffic police workers in Ahmedabad, Gujarat. *Indian Journal of Occupational and Environmental Medicine, 22*(3), 144–151.

Thaler, R., & Sunstein, C. (2008). *Nudge: Improving decisions about health, wealth, and happi-ness*. Yale University Press.

Tomkins, L., & Bristow, A. (2024). Paradoxes of organisational learning in policing: 'The truth, but not the whole truth, for everyone's sake'. *Management Learning, 55*(4), 576–595.

Tops, P. W., Bruijn, G. C. T., Spelier, R. F. J., Hogeboom, H., & Arkel, D. (2012). *The police as a learning frontline organisation*. Uitgeverij.

van Dijk, A. J., Shearing, C., & Cordner, G. (2022). Policing the pandemic: Public health, law enforcement, and the use of force. *Journal of Community Safety and Well-Being, 7*(2), 67–74.

White, R. (2013). *Environmental harm: An eco-justice perspective*. Policy Press.

White, R. (2022). Environmental crime and the harm prevention criminalist. *Frontiers in Conservation Science, 3*. https://doi.org/10.3389/fcosc.2022.1049160

Chapter 9
Eco-cops

The term 'eco-cop' was first introduced in the policing context by Philip Arthur Njuguna Mwanika (2010) in a scholarly publication, where it was used to examine the conditions for environmental policing in Eastern Africa. This book seeks to expand and redefine the term beyond its original focus on enforcing environmental laws. Instead, it conceptualises eco-cops as police officers equipped with the knowledge, skills that foster pro-environmental and pro-climate behaviour, and mindset necessary to address the cascading effects of the triple planetary crisis, among them the climate crisis (UNFCCC, 2022). This perspective challenges the traditional dichotomy between humanity and nature, emphasising the role of law enforcement in fostering resilience, community safety, and adaptation in the face of climate-related challenges. While previous chapters have taken a diagnostic approach, this chapter is more aspirational, as the eco-cop concept is still evolving. It focuses on police officers as individuals, exploring their potential to play a proactive role in climate safety and security, as well as environmental justice.

The Role of Beliefs in Shaping Green Police Officers' Behaviour

While the notion of greening the police organisation has already been touched upon in the previous chapter, there is still a dearth of literature and sources that analyse the concept of green behaviour among police officers. The scarcity of literature that exists suggests that green (police) behaviour can be understood as pro-environmental conduct or demeanour that encompasses the promotion of sustainability awareness and the benefits of ecologically friendly choices in everyday life. This, in turn, contributes to the emergence of green culture within an organisation (Konyk, 2018; Romualdo, 2022).

© The Author(s) 2025
A. Matczak, *Adapting to Climate Change in Modern Policing*, SpringerBriefs in Criminology, https://doi.org/10.1007/978-3-031-97510-3_9

The concept of pro-environmental behaviour (PEB), or green behaviour, has already been extensively researched in other disciplines and it is believed to be influenced not only by statutory controls and social norms but also by the personal beliefs and values held by individuals (Varela-Candiamo et al., 2018). Efforts to address and influence responses to climate change have been explored through both top-down structural changes (e.g. policy, regulation, and investment) and bottom-up interventions (e.g. individuals' and groups' beliefs and behaviours). Targeting individual-level behaviours is essential to complement structural changes, however, framing climate change solely as an individual-level problem is risky and can backfire, potentially causing feelings of helplessness and isolation (Creutzig et al., 2022).

A recent large-scale global study, involving 59,440 participants from 63 countries, highlights the importance of tailoring behavioural climate interventions to the audience's characteristics, specific targeted behaviours, and cultural context, with nationality and cultural factors playing a significant role (Vlasceanu et al., 2024). This mega-study used an experimental design to test the effectiveness of 11 theoretically derived behavioural interventions (outlined in the Table 9.1 below) in promoting pro-climate beliefs and behaviours. The interventions targeted outcomes such as climate-change beliefs, policy support, willingness to share information, and tree-planting contributions.

What do the study's findings mean for the police? The police, as a large organisation comprised of diverse teams and publics, face unique challenges in addressing climate change. Police officers' ability to adapt their behaviour towards climate change, like with any other sector, will likely depend on their pre-existing beliefs, whether they are sceptics, uncertain, or believers. For instance, the aforementioned study indicates that interventions, such as policy support initiatives, were generally effective only among those with high initial levels of belief in climate change, while interventions using negative emotions tended to backfire among sceptics.

Therefore, any pro-environmental behavioural change among individual police officers must be preceded by diagnostic efforts to assess their attitudes towards climate change. Significant progress has been made in other disciplines in profiling individuals' perspectives on this issue. For instance, the study of Global Warming's Six Americas does not treat the public as a homogeneous group, which is an old paradigm in attitudinal research (see Matczak, 2024); instead, it segments Americans into six distinct categories based on their beliefs, attitudes, and behaviours regarding climate change: (1) Alarmed, (2) Concerned, (3) Cautious, (4) Disengaged, (5) Doubtful, and (6) Dismissive (Leiserowitz et al., 2021). This robust, longitudinal study, conducted regularly since 2008, not only captures shifts in public opinion but also facilitates the development of tailored communication strategies aligned with each profile (ibid.). Although research on police officers' attitudes towards climate change has not yet gained significant traction, it holds potential as an important area of study in the future.

Table 9.1 Behavioural interventions in promoting pro-climate beliefs and behaviours

Intervention	Theoretical framework	Description
Dynamic social norms	Sparkman and Walton (70)	Informs participants of how country-level norms are changing and 'more and more people are becoming concerned about climate change', suggesting that people should take action
Work together norm	Howe et al. (11)	Combines referencing a social norm (e.g. 'a majority of people are taking steps to reduce their carbon footprint') with an invitation to 'join in' and work together with fellow citizens towards this common goal
Effective collective action	Goldenberg et al. (61); Lizzio-Wilson et al. (77)	Features examples of successful collective action that have had meaningful effects on climate policies (e.g. protests) or have solved past global issues (e.g. the restoration of the ozone layer)
Psychological distance	Jones et al. (10)	Frames climate change as a proximal risk using examples of recent natural disasters caused by climate change in each participant's nation and prompts them to write about the climate impacts on their community
System justification	Feygina et al. (58)	Frames climate change as threatening to the way of life in each participant's nation and makes an appeal to climate action as the patriotic response
Future self-continuity	Hershfield et al. (63)	Emphasises future self-continuity by asking each participant to project themselves into the future and write a letter addressed
		to themselves in the present, describing the actions they would have wanted to take regarding climate change
Negative emotions	Chapman et al. (45)	Exposes participants to ecologically valid scientific facts regarding the impacts of climate change framed in a 'doom and gloom' style of messaging that were drawn from different real-world news and media sources
Pluralistic ignorance	Geiger and Swim (71)	Presents real public opinion data collected by the United Nations that show what percentage of people in each participant's country agree that climate change is a global emergency
Letter to future generation	Shrum (73); Wickersham et al. (74)	Emphasises how one's current actions affect future generations by asking participants to write a letter to a socially close child who will read it in 25 years when they are an adult, describing current actions towards ensuring a habitable planet
Binding moral foundations	Wolsko et al. (59)	Invokes authority (e.g. 'from scientists to experts in the military, there is near universal agreement'), purity (e.g. keep our air, water, and land pure), and in-group loyalty (e.g. 'it is the American solution') moral foundations

(continued)

Table 9.1 (continued)

Intervention	Theoretical framework	Description
Scientific consensus	van der Linden et al. (56); van der Linden (78); Rode et al. (66)	Informs participants that '99% of expert climate scientists agree that Earth is warming and climate change is happening, mainly because of human activity'

Source: Vlasceanu et al. (2024)

Green Police Education

There is little novelty in stating that green behaviour must be adequately informed by a learning process. While it is widely accepted that awareness, attitudes, intentions, motivation, environmental education, and social norms are the primary predictors of pro-environmental behaviour, a Spanish study suggests that environmental education and internal psychological characteristics, processes, and attributes, those that shape an individual's thoughts, feelings, and behaviours, are the key drivers in predicting green behaviour (Varela-Candiamo et al., 2018).

Police education and training is, and will increasingly become, a critical factor in shaping police officers' attitudes and behaviours regarding climate change. However, this goes beyond merely updating police curricula to include units or courses on sustainability, environmental crimes, safety, and security. As highlighted by Alexandra Jones, it requires an educational approach that conveys the complexity and interconnectedness of the changes ahead. It also involves equipping officers with the skills to understand these challenges and effectively communicate them to the public during police-community interactions. The interviewee gave an example of the growing energy demands of emerging technologies, particularly cryptocurrencies and quantum computing. It was acknowledged that these advancements are shaping the future but they also represent a significant obstacle of their high energy consumption. They raise a critical question about prioritisation. If a substantial portion of energy production is allocated to these technologies, it could lead to power shortages in certain areas, with everyday consumers bearing the brunt of the consequences. How will then the governments, the police explain the power cuts to citizens? The mention of scheduled outages, even during off-peak hours, suggests a deeper concern about equity in energy distribution. This reflects a broader debate about technological progress versus resource allocation. How do we balance innovation with basic human needs, and who ultimately bears the cost of these decisions? These are the questions that should inform the police education and training as well.

This aligns with how green education is understood in the scholarly literature. Environmental education encompasses various elements, not only raising awareness or building knowledge and attitudes towards environmental challenges, but also developing skills to identify and address these issues, and encouraging active participation in resolving them. The purpose of green education is to equip individuals with the ability to critically evaluate different perspectives on an issue and enhance their problem-solving and decision-making skills (Varela-Candiamo et al., 2018).

Despite significant progress being made in this regard, police officers still lack the understanding, skills, and experience necessary for effective inspection, investigation, and multi-agency collaboration in response to environmental issues. Environmental education in police education and training remains limited in many countries. According to one of my interviewees from EUROPOL 'specialisation in environmental law enforcement is the most important issue that the police are facing (…). No dedicated and specialised unit in member states, means no investigations in Europe. No investigations in Europol means no reflection on how to do it. No reflection means no recommendation to the commission of the priority. It's like a sort of vicious circle' (Interviewee 11). The need for specialisation, as well as the establishment of dedicated green police units, was also acknowledged in the 2022 Interpol Background paper. So far, this is the only analytical document on the impact of climate change on police work published by a police organisation. The document serves as a valuable and comprehensive resource for redesigning police education curricula (Interpol, 2022).

Environmental and climate law and regulation are extremely dynamic fields, which often leads to confusion what is no longer legal. A study from Thailand revealed that while most Thai police officers were knowledgeable and skilled to get involved in environmental law enforcement, they still lacked a clear understanding or practical application of their roles and responsibilities in this area. Additionally, the proliferation of environmental legislation often lacks unity and harmonisation with other legal acts and is sometimes redundant with policies and practices carried out in other police units or external agencies (Bovornkijprasert & Rawang, 2016).

At this stage, it is almost impossible to imagine teaching about environmental matters without referring to the triple planetary crisis, consisting of air pollution, biodiversity loss, and climate change (UNFCCC, 2022). The objective of effective environmental and climate change education should be to help police officers understand the scientific and social implications of these forces on their work and lives, preparing them for an uncertain future while improving their knowledge and skills (Stevenson et al., 2017). A systematic review of literature suggests that such education should focus on making climate change information personally relevant and meaningful for police officers, with activities or educational interventions designed to engage learners (Monroe et al., 2017). This finding strongly resonates with an observation made by a senior security expert in the Netherlands (Interviewee 12), as discussed and presented earlier in the book.

Green Skillset: The Police Officers as Eco-coordinators

Climate change, among other pressing environmental issues, also impacts on the workforce development and evolving needs for specific knowledge, skills, and competencies (White, 2022). The global emergence of compliance officers, heat action planners and officers signals the changes to the workforce driven by the climate crisis. Among the many recommendations for the police to prepare for the consequences of climate changes, some expert interviewees emphasised the need for a specific skill set. The horizontal and vertical dimensions of the green criminal justice chain would necessitate that eco-cops possess excellent coordination and networking skills, combined with the ability to gather, synthesise, and communicate information and data to relevant stakeholders. This includes knowing how to assess the quality and validity of information, how to obtain the necessary information to understand the broader context, and identifying who else in the network of partners might benefit from access to this information. It also involves the ability to recognise and address environmental harms during investigations that may appear unrelated, such as those focused on drug trafficking. These skills are becoming increasingly essential as climate change denial, misinformation, and conspiracy theories continue to spread.

In an interview with Gorazd Meško, the importance of a coordinating role during times of crisis emerged from a research project examining the first major terrorist attacks in Europe, highlighting the critical need for effective crisis management and coordination. These skills and competencies are often undervalued, underfunded, and addressed only superficially within most curricula, until major shocks occur that expose their critical importance.

> Coordination is extremely important. We learned that from our study on terrorist attacks in Europe. We made a study on the aftermath of the terrorist attacks in Madrid, Paris, Brussels, London, and so on. The main problem was that the people who had to coordinate these activities were not well-prepared. It was a surprise. Those people were actually ready to work, if I may say so, in the peacetime, not in the times of crisis. It's not just about pollution or some physical processes, but it's also about human communication. (Gorazd Meško)

Many ideas and recommendations that were shared during the expert interviews resonate also with Rob White's proposal of the harm prevention criminalist—up and coming profile of a practitioner at the heart of which is the skills of brokerage: 'this refers to the ability to know who to link up with whom, which knowledge and techniques to be deployed in which circumstance, and how to maximise the effective use of material and human resources within existing fiscal limits and community settings' (White, 2022:9). In sum, fostering an individual eco-mindset within the police requires integrating green education and behaviour with the development of a complementary green skillset. Among these, the ability to coordinate environmental and climate-related matters, both vertically within the criminal justice chain or horizontally across the complex landscape of plural policing networks, is identified here as one of the key competencies that warrants greater emphasis within police education and training curricula.

Climate-Adjusted Police Work Environment

For police officers to truly develop an eco-mindset, it is not only essential to provide appropriate education and cultivate relevant skillsets; equal attention must be given to their physical preparedness and to how the physical police environment reflects and responds to broader environmental changes. Police involvement in disaster relief and management may require police officers to perform roles that are distinct from their typical responsibilities and might test unusual skills such as swimming and jumping. Alexandra Jones noted that these practical, on-the-ground operational activities within law enforcement are frequently overlooked in climate change discussions, which tend to focus primarily on strategic-level considerations. The importance of strong physical skills among police officers, particularly in response to the activities of climate change protesters, was also emphasised in an interview with Amir Niknam.

> There was this satire article that said the police thank Extinction Rebellion for their training, for the evacuations that they have to do in upcoming years, which is true. Maybe that will be the primary role of the police in ten years, just evacuating people from areas when it's uninhabitable for whatever reason.

Climate change may lead to an increase in the prevalence, distribution, and severity of known occupational hazards. Applebaum et al. (2016) identified the following occupational hazards as likely to rise due to climate change: heat, ozone, polycyclic aromatic hydrocarbons (PAHs), pathogenic microorganisms, vector-borne infectious agents, workplace violence, and wildfires. The majority of the rapidly growing body of research is dedicated to contextualising exposure to heat as an occupational hazard that impacts outdoor workers in general (Behrer & Park, 2017; Park et al., 2021; Ioannou et al., 2022). Specific studies examine the impact on traffic police (Raval et al., 2018: Mohamad Jamil et al., 2021), with some focusing solely on the experiences of female police personnel (Dey et al., 2020).

Prolonged, unmitigated exposure to heat can result in heat rash, heat cramps, heat syncope, heat exhaustion, and heat stroke (NIOSH, 2016). Heat exposure is also correlated with workplace accidents and injuries, chronic fatigue, reduced productivity, loss of concentration, and decreased alertness (Morrissey et al., 2021). Police officers' ability to monitor and assess their own risk of heat exposure and adequately mitigate it is essential. As a consequence, climate change, aside from its broader impacts, raises the very individual question of whether the frontline practitioners, including police officers, are fit to work under these adverse temperature conditions. The range of issues extends from the need for the right equipment, such as air-conditioned helmets, and lighter, more breathable uniforms, to the necessity of new approaches to shift planning. There are already good practices to draw inspiration from. For example, the G35 tropical fitness test is a medical examination mandated by German law for employees undertaking professional assignments in regions with specific climatic and health-related challenges. This assessment ensures that individuals are physically and medically fit to work in environments that may present unique health risks, including climatic stress.[1]

Besides that, according to Rob White, rostering, alongside recruitment, retention, and retirement, is a critical factor in enhancing the working environment for police officers. In many Western countries, there is a significant challenge not only in recruiting a sufficient number of police officers but also in retaining those who might otherwise continue their service. Additionally, an often-overlooked issue is the impact of rostering, particularly the strain that 12- or 16-h shifts place on officers' family lives. These individual challenges collectively exert a broader influence on police organisations as a whole. Undoubtedly, also some organisational issues contribute to the personal decisions of officers regarding their careers. This observation is complemented by an observation shared in the interview with a heat-crime expert from the United States:

> There is a need to have dynamic responses to staffing responses to heat events. When figuring out how many people to bring in on a particular day, you should be considering weather forecasts. Forecasting in much of Europe and in the United States is very good. It's highly accurate, especially with respect to what the maximum temperature on the day is going to be. We know that if the day is going to be really hot, crimes are going to increase both reported crimes and actually committed crimes. (Interviewee 10)

At the time of writing this book, a significant heatwave was affecting the Mediterranean region of Europe in summer 2024. However, I observed a lack of substantial public debate regarding the safety of first responders exposed to extreme heat. The importance of forecasting workplace challenges and anticipating increased hazards for police officers was emphasised in an interview with the same expert from the United States, who suggested the need to 'climate control their work environments'. While it is relatively straightforward to regulate indoor environments, such as in manufacturing, climate control is considerably more challenging for professions like policing, construction, or agriculture. Although air-conditioned vehicles can provide some relief, a substantial portion of police work is still conducted outdoors.

> In this broader space of how do we protect workers whose work tasks require them to be outside? There is definitely an ongoing debate in the US about what to do about that. Some of the options are shifting the timing of work that works particularly well for construction. You can work in the evenings or in the mornings or overnight so that you're less exposed to heat. That's harder to do with police because they have to be available all 24 hours a day. There's also been some exploration of cooling vests or technological solutions that would allow them to wear something that provides cooling. I haven't seen widespread adoption of those kinds of options yet. There are a lot of people who are working on those kinds of solutions and thinking about how you might implement them. (Interviewee 10)

The expert concluded by emphasising the significant risk associated with neglecting these workplace hazards. A shortage of police officers places additional strain on those in service, compelling them to operate under intense time pressure when making decisions and often requiring them to work alone. These factors can

---

[1] https://bcrt.de/en/travel-vaccinations/g35-fitness-for-the-tropics/

contribute to creating a ripple effect, impacting negatively the effectiveness and well-being of police officers in the field.

The physical aspects of police officers adapting to a climate change-shaped reality represent only one dimension of the issue. Interestingly, the psychological and social dimensions received relatively little attention in the expert interviews.

Susan Clayton's widely cited literature review (2020) on recent developments in psychology highlights the direct and indirect effects of climate change on individual well-being, emphasising that climate change is not only an environmental issue but also a psychological and social one. A growing body of evidence points to the emergence of climate anxiety, shaped by societal perceptions and interpretations of climate change and responses to it. This phenomenon has the potential to affect nearly everyone, regardless of their personal vulnerability or relative safety. Climate anxiety often co-occurs with emotions such as fear, anger, hopelessness, and grief, creating a tension between its motivating and paralysing effects. It is increasingly recognised as an ongoing and dynamically evolving environmental stressor, heightened by feelings of uncertainty and the inability to predict its precise impacts in specific locations and times. This concept resonates with Clifford Shearing's analogy, discussed in Chap. 2, comparing the current era to throwing a Lego piece into the air, symbolising the unpredictability of what comes next for all of us. Conditions like climate anxiety, along with other mental health challenges associated with environmental damage and change, such as solastalgia, eco-anxiety, eco-paralysis, and eco-nostalgia (Albrecht, 2011), are likely to increasingly affect communities, including individual police officers and their close ones.

The police operate in a unique occupational environment where mental health challenges manifest and are addressed differently than in other professions. While the stigma surrounding mental health is not as strong as it once was, underreporting remains prevalent. There is also significant ambivalence about utilising counselling services, largely due to concerns that doing so could hinder career progression (Demou, Hale & Hunt, 2020). Consequently, any future intervention strategy aimed at addressing the impact of climate change on police officers' mental health must take into account the distinct characteristics and cultural nuances of the police culture.

Recruitment of New Generations

One of the most pressing organisational challenges faced by police forces in many Western countries is staff shortages, difficulties in retaining personnel, and challenges in recruiting sufficient numbers of new officers. In several expert interviews, concerns were raised about the police becoming an increasingly unattractive employer for younger generations. These young individuals, who are often more concerned about environmental issues and more vulnerable to climate anxiety (Clayton, 2020), may perceive the police as a necessary but problematic institution, particularly due to their involvement in policing climate and environmental demonstrations. Additionally, unlike older generations, younger people may not feel the same sense of loyalty to commit to a lifelong career in the service. A police officer from the Netherlands described it as:

> In the current job market, the police are not a very attractive employer. Those who are attracted to working for the police are the people who are inherently motivated and tend to be more conservative. It's not like the old days. That's also a thing that has changed with the incoming younger generation. The loyalty is not a given. It's no longer a job for life. (Interviewee 3)

In the Netherlands, this even led to informal attempts by off-duty police officers attending climate-change demonstrations to build understanding and explore new (and informal) approaches to recruitment. Amir Niknam described one of the initiatives of bringing a recruitment banner with a QR code that directed people to the recruitment website for environmental crime units, accompanied by the catchy slogan, 'Give your green heart blue power' (*Geef je groene hart blauwe slagkracht*). While some viewed the initiative as somewhat controversial, it was widely met with enthusiasm among the protesters.

> We spoke to hundreds of people that day, and everybody was so happy to see that the police are not just a neutral group of people who are slowly walking into the demise of humanity, that there are also people within the police that give a shit. Interestingly enough, the fact that we said this is an important topic and if you want to help out, you can join us, the fact that it was our message, helped us bond with the group and helped lower the tensions that was there. I think in terms of community policing, it was a great job. (Amir Niknam)

However, as with any observation in this book, caution is advised when attempting to generalise. Evidence from other parts of the world suggests that while recruitment for police forces becomes increasingly challenging in developed countries of the Global North, climate change may paradoxically create new opportunities for recruitment into police and other public services in less developed countries. This perspective was highlighted and explained by Peter Schwartzstein, an environmental journalist and author of the book *The Heat and the Fury: On the Frontlines of Climate Violence* as follows:

> In so many parts of the world, for parts of the world in which I work, as agriculture collapses to varying degrees due to climate stresses and due to other problems, public sector employment is more attractive and more sought after than ever before.
>
> In countries like Iraq, Sudan, before the beginning of the civil war, in countries like Egypt, being a policeman or being in the military or in the security services in any capacity becomes more and more attractive, as farming becomes less and less viable. It's a job with strong job security (…). It's partly, of course, because you've got much higher population growth rates in other parts of the world. There isn't quite the same surplus of alternative jobs for people to potentially turn to in many of these communities, in the rural parts of poor countries. You also have very limited education, and hence in many instances, people don't have easily transferable skills to work in a kind of non-agricultural setting. (Peter Schwartzstein)

In sum, there are multiple profiles of eco-cops, reflecting a spectrum of motivations and degree of environmental engagement. Ideally, many would be intrinsically motivated individuals who prioritise environmental values. However, it is equally important to acknowledge the role of top-down, organisationally driven behavioural change, and the potential to influence officers by cultivating a culture of environmental awareness within the institution. Finally, it must also be recognised that some officers may be assigned to roles requiring pro-environmental and pro-climate behaviour, even if such responsibilities do not align with their personal beliefs or values.

Care and Humanitarian Mindset

In the book *Security in the Anthropocene* Cameron Harrington and Clifford Shearing offer a new conceptual framework to understand how safety and security will be practised and narrated on a planet that is increasingly affected by climate change (Harrington & Shearing, 2017). The final chapter of the book discusses the emerging ethos of care among security practitioners as a response to the climate crisis. In my conversation with Clifford Shearing, I asked him about the ending of the book and how he thinks about the viability of this option today:

> I'm actually more optimistic about it now than when we wrote that chapter because I see it happening everywhere. If you look at the centre at Leeds in York that Adam Crawford established,[2] a central thing is vulnerability policing, which is not just about policing vulnerable populations but enabling vulnerable populations to care for themselves. I'm actually seeing an ethos of care emerging. Resilience policing is an ethos of care. How do we help you care for yourself? How do we assist you in your care activities (…) A lot of the climate work was to avoid this move from the Holocene into the Anthropocene. Now that we don't seem to be doing very well with that avoidance, it's how do we live in the Anthropocene? How do we live with heat? How do we live with wind? How do we live with more and more tornadoes? There's a shift here. It is a care shift, I think. How do we care for people in those circumstances (…) that's part of a policing issue. That's a safety production issue. That's an ethics of care.

Such an ambition requires a reprioritisation of humanitarian values in police education and training. As Rob White emphasised: 'We have to elevate the levels of expertise, training, and education to effectively respond to periods of crisis because we need lateral thinking. If I were to address the police, I would stress that, to make policing more attractive, it is essential to highlight its humanitarian roles and raise overall expectations'.

Climate laws, policies, and interventions too often neglect the human factor in the unfolding climate change-related events. In criminology, the police are often depicted in a bipolar manner: either as part of the criminal justice system or as one of many security actors in the pluralised and nodal web of public, private, and civil entities. The climatisation of police and policing will increasingly pull the police in both familiar and unfamiliar directions. Nonetheless, this also means that the worsening climate situation will most likely lead to the expansion of the security sector,

---

[2] Vulnerability and Policing Futures Research Centre.

and this is something that might in fact challenge the potential and viability of the ethics of care among police officers.

Policing civil disobedience and handling environmental protests already make them appear as ruthless enforcers in the hands of the government and gatekeepers to the criminal justice system. However, the police's involvement in disaster relief and management, as emphasised by some of my interviewees, provides them with an opportunity to demonstrate their caring and humanitarian side. Especially disasters have a power to trigger a sense of solidarity in communities. However, police responses to disasters are highly context-dependent and significantly influenced by police-public dynamics. Moreover, any documented signs of solidarity are often fleeting, with public support and cooperative efforts frequently diminishing over time. When it comes to policing the crimes of powerful and climate offenders, the state police still seem to be on the standby, observing how other policing actors, under the guise of administrative and regulatory institutions, fulfil—or fail to fulfil—their tasks.

Climate crisis not only affects people's health, safety but also sense of belonging, self-esteem, self-efficacy, which might lead to something that is now called in the literature as ecological grief or eco-anxiety (Adger et al., 2022). In whatever assemblage or collaboration an eco-cop finds themselves in, the most important and time-consuming currency to make this caring work is trust. The human element is usually underestimated in joint working initiatives, whether formal or informal, and at the operational level, collaborations work best when they are based on trust (White, 2022).

The legal obligation of a duty of care has become a central tenet and a prominent feature of climate litigation initiatives, which has significantly enhanced their public standing. Adopting a care-oriented approach appears to be one of the most effective strategies for navigating the complexities of climate disruption while preserving societal legitimacy. This mindset can be interpreted as the ultimate embodiment of the eco-cop ethos. Accordingly, the police service should not be denied the opportunity to engage with and adopt this perspective. Whether the intensifying effects of climate change, alongside the expanding climatisation of police work, will succeed in fostering a more caring and humanitarian orientation among police officers remains to be seen.

References

Adger, W. N., Barnett, J., Heath, S., & Jarillo, S. (2022). Climate change affects multiple dimensions of well-being through impacts, information and policy responses *Nature Human Behaviour, 6*(11), 1465–1473.

Albrecht, G. (2011). Chronic environmental change: Emerging 'psychoterratic' syndromes. In I. Weissbecker (Ed.), *Climate change and human well-being* (pp. 43–56). Springer.

Applebaum, K. M., Graham, J., Gray, G. M., et al. (2016). An overview of occupational risks from climate change. *Current Environmental Health Reports, 3*(1), 13–22.

Behrer, A. P., & Park, J. (2017). *Will we adapt? Temperature, labour and adaptation to climate change*. Harvard Project on Climate Agreements Working Paper, 16–81. Retrieved from https://www3.nd.edu/~nmark/Climate/Park_will_we_adapt.pdf. Accessed 16.08.2024.

Bovornkijprasert, S., & Rawang, W. (2016). Integration of environmental education and environmental law enforcement for police officers. *International Journal of Environmental & Science Education, 11*(12), 5698–5709.

Clayton, S. (2020). Climate anxiety: Psychological responses to climate change. *Journal of Anxiety Disorders, 74*, 102263. https://doi.org/10.1016/j.janxdis.2020.102263

Creutzig, F., Niamir, L., Bai, X., et al. (2022). Demand-side solutions to climate change mitigation consistent with high levels of wellbeing. *Nature Climate Change, 12*, 36–46.

Dey, A., Mishra, T., Sahu, S., & Saha, A. (2020). Occupational heat exposure of female police personnel: Its implication of climate. *The Indian Police Journal, 67*(3), 52–57.

Harrington, C., & Shearing, C. (2017). *Security in the anthropocene. Reflections on safety and care*. Transcript.

Demou, E., Hale, H. & Hunt, K. (2020). Understanding the mental health and wellbeing needs of police officers and staff in Scotland. *Police Practice and Research, 21*(6), 702–716.

Interpol Innovation Centre. (2022). *Crime, law enforcement and climate change*. Interpol Innovation Centre. Background Paper. Retrieved from file:///C:/Users/annaf/Downloads/IC%20Background%20Paper%20-%20Crime%20Law%20Enforcement%20and%20Climate%20Change-4.pdf. Accessed 16.08.2024.

Ioannou, L. G., Foster, J., Morris, N. B., Piil, J. F., Havenith, G., Mekjavic, I. B., Kenny, G. P., Nybo, L., & Flouris, A. D. (2022). Occupational heat strain in outdoor workers: A comprehensive review and meta-analysis. *Temperature (Austin), 9*(1), 67–102.

Konyk, J. (2018). *Green policing: Recommended actions for an environmental sustainability plan for the Vancouver Police Department*. Retrieved from https://sustain.ubc.ca/sites/default/files/2018-54%20Green%20Policing%20Recommended%20Actions%20for%20an%20Environmental%20Sustainability%20Plan%20at%20the%20VPD_Konyk.pdf. Accessed 16.08.2024.

Leiserowitz, A., Roser-Renouf, C., Marlon, J., & Maibach, E. (2021). Global Warming's Six Americas: A review and recommendations for climate change communication. *Current Opinion in Behavioural Sciences, 42*, 97–103.

Matczak, A. (2024). Who is "the public" when we talk about crime? Interpreting and framing public voices in criminology. In K. Stockdale & M. Addison (Eds.), *Marginalised voices in criminology* (pp. 166–182). Routledge.

Mohamad Jamil, P., Karuppiah, K., Rasdi, I., How, V., Mohd Tamrin, S., Mani, K., Sambasivam, S., Naeini, H., Mohammad Yusof, N., & Hashim, N. (2021). Occupational hazard in Malaysian traffic police: Special focus on air pollutants. *Reviews on Environmental Health, 36*(2), 167–176.

Monroe, M. C., Plate, R. R., Oxarart, A., Bowers, A., & Chaves, W. A. (2017). Identifying effective climate change education strategies: A systematic review of the research. *Environmental Education Research, 25*(6), 791–812.

Morrissey, M. C., Brewer, G. J., Williams, W. J., Quinn, T., & Casa, D. J. (2021). Impact of occupational heat stress on worker productivity and economic cost. *American Journal of Industrial Medicine, 64*(12), 981–988.

Mwanika, P. A. N. (2010). *Eco-cop. Environmental policing in Eastern Africa*. Paper 215. Institute for Security Studies. Retrieved from https://www.files.ethz.ch/isn/136679/PAPER215.pdf. Accessed 16.08.2024.

National Institute for Occupational Safety and Health (NIOSH). (2016). *Criteria for a recommended standard: Occupational exposure to heat and hot environments (DHHS (NIOSH) Publication No. 2016-106)*. U.S. Department of Health and Human Services, Centers for Disease Control and Prevention. Retrieved from https://www.cdc.gov/niosh/docs/2016-106/pdfs/2016-106.pdf. Accessed 19.02.2025.

Park, R. J., Pankratz, N., & Behrer, A. P. (2021). *Temperature, workplace safety, and labor market inequality*. IZA Discussion Papers, No. 14560. Institute of Labor Economics (IZA). Retrieved from https://www.econstor.eu/bitstream/10419/245611/1/dp14560.pdf. Accessed 10.08.2024.

Raval, A., Dutta, P., Tiwari, A., Ganguly, P. S., Sathish, L. M., Mavalankar, D., & Hess, J. (2018). Effects of occupational heat exposure on traffic police workers in Ahmedabad, Gujarat. *Indian Journal of Occupational and Environmental Medicine, 22*(3), 144–151.

Romualdo, A. (2022). Prospect for green policing: Constructs and dimensions of environmental sustainability in the context of public safety. *Global Sustainability Research, 1*(2), 22–29.

Stevenson, R. B., Nicholls, J., & Whitehouse, H. (2017). What is climate change education? *Curriculum Perspectives, 37*, 67–71.

UNFCCC. (2022). *What is the triple planetary crisis?* United Nations Framework Convention on Climate Change. Retrieved from https://unfccc.int/news/what-is-the-tripleplanetary-crisis. Accessed 16.08.2024.

Varela-Candamio, L., Novo-Corti, I., & García-Álvarez, M. T. (2018). The importance of environmental education in the determinants of green behaviour: A meta-analysis approach. *Journal of Cleaner Production, 170*, 1565–1578.

Vlasceanu, M., et al. (2024). Addressing climate change with behavioural science: A global intervention tournament in 63 countries. *Science Advances, 10*(6), *5778*. DOI: 10.1126/sciadv.adj5778

White, R. (2022). Environmental crime and the harm prevention criminalist. *Frontiers in Conservation Science, 3*. https://doi.org/10.3389/fcosc.2022.1049160

Chapter 10
Conclusions

The impact of climate change respects neither geographical boundaries nor convenient timing, with its consequences increasingly and visibly permeating all spheres of life. The fingerprints of climate change, shaped by local circumstances and conditions, manifest somewhat differently across regions worldwide. The common thread is that the combination of these accelerating effects and the painfully slow progress in mitigating them resembles a slow death from carbon monoxide poisoning in any part of the world. In the future, country-, region-, and sector-specific case studies, as well as analyses of weather phenomena, disaster types, or climate-related crimes, will help narrow down the topic of this book and provide detailed insights into broader themes explored in this work.

Climate change is increasingly shaping the realities of police work, often subtly and without clear attribution, through a series of disruptive shocks and stressors. This process manifests in the form of 'pain distribution', that pulls the police in many different directions than what they have traditionally known. Whether or not police institutions recognise it, their practices are becoming climatised, either by adapting to the impacts of climate change or by integrating climate-related policies. These shifts may be symbolic, strategic, precautionary, or transformative, but they are never static. Shaped by both internal and external factors, these processes might advance or regress over time. What remains clear is that police and policing will not remain unchanged; it will evolve, but it is not certain that this will always happen in ways that align with pro-environmental or climate-conscious ideals.

Heightened recognition, acknowledgement, and proactive response to these changes, are largely missing in this field. The pace of change within police organisations remains insufficient given the scale of the challenge. While collecting data and working on this book, comments were made suggesting that this topic is uncharted territory, our knowledge is limited, and the uncertainty and unpredictability of what lies ahead make preparation difficult. It is true that our understanding of how the cascading effects of climate change will specifically impact police and policing still remains limited and perhaps even distorted. Furthermore, the growth of green

A. Matczak, *Adapting to Climate Change in Modern Policing*, SpringerBriefs in
Criminology, https://doi.org/10.1007/978-3-031-97510-3_10

criminology and the expansion of environmental legislation have largely overlooked the role of the police in these processes. Nevertheless, as in other domains, the increasing recognition of the pivotal relationship between humanity and nature will inevitably push criminology and policing studies to revisit their respective fields and redefine their roles in this rapidly changing world.

Using Clifford Shearing's metaphor, the police, like all other social institutions, are being tossed into the air like Lego pieces, and upon landing, they face a significant restructuring of the crime and harm landscape. One of the first questions that comes up in meetings with police officers concerns their evolving job portfolio and which societal issues will become their responsibility, or 'police property'. A key area of interest for the police is the changing spectrum of violations they will, or might, be tasked with monitoring, policing, and investigating. Although not always explicitly linked to climate change, the increasing recognition of environmental harms, the significant expansion of environmental legislation, and the emergence of specialised environmental crime units within police forces already point to a broadening of their portfolio. The ecocide movement, by fostering debate, has further heightened global sensitivity to environmental issues and shifted attention towards crimes often perpetrated by powerful actors. Recent successes in climate litigation have also contributed to reimagining not only the profiles of offenders but also the concept of victimhood. These developments lay important groundwork for advancing the green transition and reset within policing and encouraging the adoption of an eco-cop mindset. Such a shift could gradually increase police awareness of other aspects of climatisation in their work, paving the way for more environmentally responsive policing practices.

Aside from the increasing frequency of disasters, climate change will most often manifest through higher temperatures, which, in turn, will destabilise the availability of essential resources and the conditions necessary for survival. Unsurprisingly, this topic is rapidly gaining significant attention, particularly in discussions framed under the umbrella terms of climate and security, climate and conflict, and climate and violence. There is a growing body of scholarship exploring the intersection between heatwaves and human behaviour. A subset of this research is already investigating the relationship between high temperatures and individuals' propensity for conflict and crime. Evidence increasingly demonstrates that higher temperatures and other consequences of climate change act as amplifiers, driving population movement within and across borders. This, in turn, might be becoming a factor in the rise of hate crimes, youth radicalisation, and extremism. However, this emerging area of scholarship has thus far seen minimal involvement from criminologists. For many in the field, the predominant frameworks for interpreting criminality remain rooted in sociological perspectives and the social construction of crime, leaving a gap in integrating climate-related dynamics into criminological research.

Despite the many ways in which climatisation processes shape police work, it is the subject of climate and environmental demonstrations, or more specifically, police responses to these protests, that brings the nexus of climate change and policing to the forefront at this moment. This is because the police are often seen as the agency tasked with 'securing a social order that is disintegrating', as one of my interviewees aptly observed (Interviewee 15). Policing civil disobedience in this context appears to operate on at least three distinct levels. The first, and the one that has received the most criminological attention, involves the policing of environmental and/or climate demonstrations. Equally resource-intensive, however, are protests against perceived greenwashing, commonly referred to as farmers' protests, which take on culturally specific forms in different countries. The third, and least explored, aspect of policing civil disobedience involves direct or indirect public outrage, protests, and hate speech directed at incoming migrants, including climate refugees. This complex area of research, policy, and police practice is inevitably interwoven with critical issues such as police discretion, accountability, procedural justice, and public attitudes towards police legitimacy. While this intersection of climate change and police work is just one of many aspects, it is the area that garners the most attention in criminological and public debates. It also strains the image of the police, often portraying them as oppressors, particularly when their actions are perceived as suppressing dissent rather than maintaining order.

On the other hand, the subject of policing disasters draws the police not only into a broad network of security and emergency partners but also provides an opportunity for them to develop and embrace a humanitarian mindset. Much of this work will take the form of incidental activities, for which proper preparation will become increasingly essential. This intersection of climate change and police work also raises critical questions about the evolving crime and harm landscape. It underscores the importance of understanding disasters through their three distinct phases: pre-disaster, disaster, and post-disaster. Such an approach can help police organisations address the unique public safety requirements and violations associated with each phase. Furthermore, this area of police activity is likely to highlight the concept of plural policing, where the ability to collaborate effectively with a diverse range of partners becomes a highly valuable skill.

Contrary to popular opinion, there are examples of police services that have already recognised certain aspects of climate change impacts and are beginning to adapt. This is a slow and challenging process, where the police may be perceived as learning with great difficulty. However, it must be acknowledged that only recently have police services been tasked with both 'going green' and 'protecting the green' simultaneously. The introduction of sustainability policies and practices, the establishment of environmental crime units, and the development of national and international collaborations to enhance capacity for environmental crime enforcement are some examples of this emerging trend in certain countries. This reflects a reconfiguration—and sometimes a contestation—of everyday policing as the police adapt on the go. While the future appears uncertain, the police have the opportunity to view this challenging scenario as a catalyst for growth and transformation.

Some police services have sought advice from the military, a move that has been interpreted by some as risky, as it could lead to greater securitisation practices and intensify the framing of environmental issues as threats. Nevertheless, valuable lessons can be drawn from the military sector, such as the importance of avoiding surprises and scenario-based training framework. The concept of resilience or adaptive policing suggests that a growth mindset within the police is essential, challenging the notion that police organisations are resistant to learning. Although evidence of siloed working practices persists, there is a gradual recognition of the need for a paradigm shift, emphasising enhanced collaboration with other sectors. The pace and depth of this learning process within police organisations may be shaped by the extent and intensity of lived experiences of climate change consequences. However, it is increasingly clear that traditional methods of policing are no longer sustainable. A resilient, growth-oriented framework highlights that linear policy and practice development is unrealistic and that adaptive, innovative approaches will be critical for the future of policing in a changing climate.

The police are made up of individual people, and I have had the opportunity to meet and speak with many who already embody the eco-cop mindset. Thus, it is hoped that this book will inspire a rise in both the quantity and quality of environmentally minded and climate-conscious police officers around the world—officers who are experts in their field, have access to specialised and multidisciplinary training, work in a climate-adjusted work environments, and embody environmentally responsible behaviours. These officers will need to excel in their coordinating roles, regardless of the collaborations they participate in, and care not only about the safety of their communities but also about the well-being of the planet, also a victim of climate change crimes. For the climatisation of police and policing to be noticeable, it needs to become personal for police officers, enabling them to fully embrace and adopt an eco-mindset. While it may be unrealistic to expect all officers to adopt this perspective, there must be a critical mass within police organisations who recognise and acknowledge the impending impacts of climate change on their work. This shift is essential for the future of environmentally responsive policing.

This book might be guilty of many omissions, considering the breadth and magnitude of challenges the police will face as the effects of climate change accelerate. Despite efforts to include insights from experts across the globe, the book remains largely written from a Global North and European perspective. It aims to foster a deeper understanding of the police's role in addressing the implications of climate change, blending current debates in criminology and policing studies with practical examples that serve as a sort of climate change handbook for police practitioners. While progress is being made in some areas, no one is moving quickly enough. As one of the experts aptly put it: 'I always cite Eisenhower when he talked about his struggle between the urgent and the important. The urgent is the day-to-day business, and the important is tomorrow's business. Climate is tomorrow's business. What we see on the political level, but also in organisations, is that we are so busy with the daily business that everything that is more long-term is kind of a sideshow' (Interviewee 12). This sentiment highlights the critical need for a shift in focus to prioritise the implications of climate change for police practice, and it is hoped that this book will be of service to those who share the same sense of urgency and importance.

Index

© The Author(s) 2025
A. Matczak, *Adapting to Climate Change in Modern Policing*, SpringerBriefs in
Criminology, https://doi.org/10.1007/978-3-031-97510-3

The manufacturer's authorised representative in the EU is Springer
Nature Customer Service Centre GmbH, Europaplatz 3, 69115 Heidelberg,
Germany. If you have any concerns regarding our products, please
contact ProductSafety@springernature.com

Printed and bound by CPI Group (UK) Ltd, Croydon, CR0 4YY

24/04/2026

02096374-0003